T0139784

Biogeography-Based Optimization: Algorithms and Applications

Yujun Zheng · Xueqin Lu
Minxia Zhang · Shengyong Chen

Biogeography-Based Optimization: Algorithms and Applications

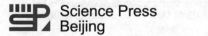
Science Press
Beijing

Springer

Yujun Zheng
Hangzhou Institute of Service Engineering
Hangzhou Normal University
Hangzhou, Zhejiang, China

Minxia Zhang
College of Computer Science
and Technology
Zhejiang University of Technology
Hangzhou, Zhejiang, China

Xueqin Lu
College of Computer Science
and Technology
Zhejiang University of Technology
Hangzhou, Zhejiang, China

Shengyong Chen
College of Computer Science
and Technology
Zhejiang University of Technology
Hangzhou, Zhejiang, China

ISBN 978-981-13-4794-8   ISBN 978-981-13-2586-1   (eBook)
https://doi.org/10.1007/978-981-13-2586-1

Jointly Published with Science Press, Beijing, China

The print edition is not for sale in China Mainland. Customers from China Mainland please order the print book from: Science Press.

This Springer imprint is published by the registered company Springer Nature Singapore Pte Ltd.
The registered company address is: 152 Beach Road, #21-01/04 Gateway East, Singapore 189721, Singapore

Preface

Nature-inspired optimization algorithms are a class of heuristic algorithms that mimic natural phenomena to solve complex optimization problems which are intractable by traditional exact algorithms. Since John Holland developed the genetic algorithm (GA) in the early 1970s, nature-inspired algorithms have aroused great research interests. The last decade has witnessed a rapid development of new nature-inspired algorithms which exhibit wide applicability and promising performance on a variety of engineering problems.

In 2008, Dan Simon proposed a new algorithm named biogeography-based optimization (BBO), which mimics the evolution process of habitats based on migration to evolve a population of solutions toward the global optimum or a near-optimum. BBO exhibits competitive performance compared with many popular algorithms and has attracted much interest from academia and industry.

Our research team has been working on nature-inspired algorithms and their applications for a decade. Since the proposal of BBO, we have done a lot of work on the improvement of BBO as well as the application of BBO to many engineering optimization problems. The results of our work have been published in well-known academic journals and conferences including Computer & Operations Research, IEEE Transactions on Evolutionary Computation, IEEE Transactions on Fuzzy Systems, IEEE Transactions on Intelligent Systems, Applied Soft Computing, IEEE Congress on Evolutionary Computation (CEC), International Conference on Swarm Intelligence. In particular, in 2014 we proposed a major improvement of BBO, called ecogeography-based optimization (EBO). Also in 2014, our study on the application of BBO to the emergency scheduling of engineering rescue tasks in disasters won the Runner-Up of International Federation of Operational Research Societies (IFOPS) Prize for Development.

This book is primarily intended for researchers, engineers, and students who are interested in BBO and/or who want to borrow ideas from BBO in their researches in all aspects of intelligent computing. The chapters cover the basic BBO algorithm and its variants, extensions, and applications. Chapter 1 introduces the basic concepts of optimization problems and algorithms, Chap. 2 gives a brief overview of the BBO algorithm and its developments, Chaps. 3 and 4 present two important improved versions of BBO, Chap. 5 describes hybrid BBO algorithms, and Chaps. 6–9 describe the application of BBO to optimization problems in four typical areas including transportation, job scheduling, image processing, and machine learning. A part of source code and data sets used in this book can be found in http://compintell.cn/en/dataAndCode.html.

Besides the authors, our students including Xiaohan Zhou, Yichen Du, Bei Zhang, and Xiaobei Wu have also made essential contributions in algorithm testing, result analysis, and proofreading of the manuscript. We are grateful for the help of Prof. Haiping Ma of Shaoxing University and Prof. Qi Xie of Hangzhou Normal University. We would also like to thank Mr. Yingbiao Zhu of the Science Press, China. The work was supported by National Natural Science Foundation of China (Grant Nos. 61473263 and U1509207), Zhejiang Provincial Natural Science Foundation (Grant No. LY14F030011), and Scientific Research Starting Foundation of Hangzhou Normal University (Grant No. 4115C5021820424).

Hangzhou, China   Yujun Zheng
June 2018   Hangzhou Normal University

Contents

Chapter 1
Optimization Problems and Algorithms

Abstract Optimization problems occur in almost everywhere of our society. According to the form of solution spaces, optimization problems can be classified into continuous optimization problems and combinatorial optimization problems. Algorithms for optimization problems, according to whether they can guarantee the exact optimal solutions, can be classified into exact algorithms and heuristic algorithms. This chapter presents a brief overview of optimization problems and then introduces some well-known optimization algorithms, which lays the foundation of this book.

1.1 Introduction

Optimization is common in almost all spheres of human activities, including labor division and cooperation in caveman hunting, intensive farming in agricultural production, job scheduling in industrial manufacturing, nutrition mixture in daily diet, design and implementation of national policies. In the early days of human being's history, optimization is mainly based on personal experiences. With the development of mathematics, more and more optimization is realized by using accurate mathematical methods. Since the end of the twentieth century, fast development of computer and artificial intelligence technologies has provided new effective and efficient methods to solve complex optimization problems that were heretofore considered to be unsolvable or unaffordable.

When investigating optimization methods, scientists are often inspired by nature which it is itself a "great optimizer." For example, in species evolution, species that did not adapt to the environment were gradually eliminated; meanwhile, the others continually improve their competitiveness by trying to optimally combine good genes. Inspired by this, Holland [9] proposed genetic algorithms (GA) to solve the optimization problem. In metal cooling, all molecules try to move to the lowest energy state and thus are ultimately arranged in order. Inspired by this, Kirkpatrick [12] proposed simulated annealing (SA) algorithm. A small ant's strength is very little, but the ant colony can organize its members in a fantastic way to accomplish complicated tasks such as foraging and nesting efficiently. Inspired by this, Dorigo [4] proposed ant colony optimization (ACO) algorithm. Similarly, Kennedy and Eberhart [10] proposed particle swarm optimization (PSO) algorithm based on the social

© Springer Nature Singapore Pte Ltd. and Science Press, Beijing 2019
Y. Zheng et al., *Biogeography-Based Optimization: Algorithms and Applications*,
https://doi.org/10.1007/978-981-13-2586-1_1

behavior of bird flocking, and Karaboga [2] proposed artificial bee colony (ABC) algorithm based on the self-organized behavior of bee colonies. There are also many other nature-inspired algorithms, such as differential evolution (DE) [23], harmony search (HS) [7], fireworks algorithm (FWA) [24], gravitational search algorithm (GSA) [19], water wave optimization (WWO) [29].

In recent years, a new algorithm named biogeography-based optimization (BBO) [21] has achieved great success. BBO solves optimization problems by mimicking species migration among habitats in biogeography. It is theoretically sound, simple, and efficient and thus has attracted a large number of researchers and has been applied to a wide range of engineering optimization problems.

In this chapter, Sect. 1.1 introduces the basic concepts of optimization problems, Sect. 1.2 presents some traditional exact optimization algorithms, and Sect. 1.3 presents some typical heuristic algorithms including GA and SA. From the next chapter, we will introduce BBO algorithms and their applications.

1.2   Optimization Problems

An optimization problem is to find the best solution among a set of feasible candidates. But how can we say that a solution is the "best" or is "better" than another? It is determined by the *objective function* of the problem. Suppose that the minimum objective function value is expected, the problem can be formulated as:

$$\begin{aligned} \min \quad & f(\mathbf{x}) \\ \text{s.t.} \quad & g(\mathbf{x}) \geq 0 \\ & \mathbf{x} \in \mathbb{D} \end{aligned} \qquad (1.1)$$

where $\mathbf{x}$ is the decision variable (sometimes a scalar, but in most times a vector), $\mathbb{D}$ is the domain of $\mathbf{x}$, $f : \mathbb{D} \to \mathbb{R}$ is the objective function, and g is the set of constraint functions.

Equation (1.1) is a general form. The problem to maximize $f(\mathbf{x})$ can be transformed to an equivalent problem that minimizes $-f(\mathbf{x})$. Equality or less-than inequality constraints can be transformed to greater-than-or-equal-to inequality constraints. An optimization problem is called a constrained optimization problem if the cardinality of constraint set g is larger than zero, otherwise is called an unconstrained one.

According to the distribution of solutions in $\mathbb{D}$, optimization problems can be divided into continuous optimization problems and combinatorial optimization problems.

1.2.1   Continuous Optimization Problems

As the name suggests, continuous optimization problems have continuous solution spaces. One of its most common forms is to find the optimal value of a continuous real function in a given range, e.g.,

$$\min \quad 4x^3 + 15x^2 - 18x$$
$$\text{s.t.} \quad -1 \le x \le 1 \tag{1.2}$$

The problem shown in Eq. (1.2) is a one-dimensional optimization problem; i.e., it has only one decision variable. Equation (1.3) gives a multi-dimensional optimization problem with a more complex objective function:

$$\min \quad \sum_{i=1}^{n} \sum_{k=0}^{20} \left(0.5^k \cos(3^k \pi x_i + 0.5) + \sin(5^k \pi x_i)\right)$$
$$\text{s.t.} \quad -100 \le x_i \le 100, \quad i = 1, 2, \ldots, n \tag{1.3}$$

where n denotes the number of dimensions. Obviously, the higher the dimension, the larger the solution space is, and the more difficult it is to solve the problem.

Next let us see an example of constrained continuous optimization problems. Suppose that every day a person needs at least 100 g carbohydrate and at least 75 g protein, but cannot intake more than 90 g fat. There are only two types of food, named A and B, whose prices are \$12/kg and \$16/kg, respectively, and their nutrimental compositions are shown in Table 1.1. The problem is to determine the quantities of food A and B, such that the nutritional requirements of the person are satisfied and the total cost is minimized.

Let x_1 and x_2 denote the quantities (in kg) of A and B, according to the objective of cost minimization and the constraints on nutritional requirements, the problem can be formulated as Eq. (1.4):

$$\min \quad 12x_1 + 16x_2$$
$$\text{s.t.} \quad 125x_1 + 100x_2 \ge 100$$
$$60x_1 + 120x_2 \ge 75$$
$$60x_1 + 72x_2 \le 90 \tag{1.4}$$
$$x_1 \ge 0, x_2 \ge 0$$

Let us see another problem related to a sensor network. The network already has three sensors named A, B, and C, and now, we need to add a new sensor denoted as S, as illustrated in Fig. 1.1, where the circle in shallow denotes a pond in which the sensor cannot be placed. The center of the circle is located at (18,25), and the radius of the circle is 5. The locations of the three existing sensors as well as their daily hours for communicating with S are given in Table 1.2. The energy consumption of communication between two sensors is $2d^{1.6}$ per hour, where d is the distance

Table 1.1 Nutrimental compositions (in grams) of food A and B

Food	carbohydrate	protein	fat
A	125	60	60
B	100	120	72

Fig. 1.1 Illustration of the sensor network

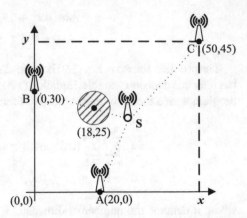

Table 1.2 Locations of the three existing sensors and their hours for communication with the new sensor S

Sensor	Location	Communication hours
A	(20,0)	2.5
B	(0,30)	3.5
C	(50,45)	3

between the two sensors. The problem is to determine the location (x, y) of the new sensor S, such that the total energy consumption of communication to S is minimized. The constraint is that S cannot be placed in the pond. Thus the problem can be formulated as Eq. (1.5):

$$
\min \ 5\left(\sqrt{(x-20)^2+y^2}\right)^{1.6} + 7\left(\sqrt{x^2+(y-30)^2}\right)^{1.6} + 6\left(\sqrt{(x-50)^2+(y-45)^2}\right)^{1.6}
$$

$$
\text{s.t.} \ \sqrt{(x-18)^2+(y-25)^2} \geq 5 \tag{1.5}
$$
$$
0 \leq x_1 \leq 50, 0 \leq x_2 \leq 45
$$

A great challenge to solve many continuous optimization problems, in particular those with complex objective functions, is the existence of local optima. A local optimum is a solution that is the best among its neighborhood, but not as good as the global optimum in the whole solution space. Figure 1.2 illustrates the global optimum and local optima of a one-dimensional continuous optimization problem. Not all the optimization algorithms can guarantee to find the global optimum for every problem—sometimes they can just find local optima. Some local optima are only a little worse than the global optimum, while some can be much worse, and the latter is that we should try to avoid.

Fig. 1.2 Illustration of global optimum and local optima

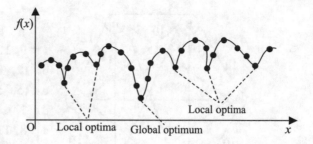

1.2.2 Combinatorial Optimization Problems

Unlike continuous optimization problems, combinatorial optimization problems have discrete solution spaces. The word "combinatorial" refers to the fact that such problems often consider the selection, division, and/or permutation of discrete components. The 0–1 knapsack problem is one of the most famous combinatorial optimization problems. The problem is to choose a subset of n items, each with a size a_i and a value c_i $(1 \leq i \leq n)$, such that the total value of the subset is maximized without having the size sum to exceed to knapsack capacity b. The problem can be formulated as Eq. (1.6):

$$\max \quad \sum_{i=1}^{n} c_i x_i$$

$$\text{s.t.} \quad \sum_{i=1}^{n} a_i x_i \leq b \tag{1.6}$$

$$x_i \in \{0, 1\}, \quad i = 1, 2, \ldots, n$$

where $x_i = 1$ denotes that the ith item is chosen into the knapsack and $x_i = 0$ otherwise. The problem's domain $\mathbb{D} = \{0, 1\}^n$ is discrete, and the problems on such domain are collectively called 0–1 programming problems.

If there are n types of items and we can choose one or more items of each type, then the problem extends to the integer knapsack problem, whose formulation just replaces $x_i \in \{0, 1\}$ with $x_i \in \mathbb{Z}^+$ in Eq. (1.6); i.e., each decision variable x_i becomes an integer. Such problems are collectively called integer programming problems.

The knapsack problem focuses on the "selection" of components. A typical problem concerning the "permutation" of components is the traveling salesman problem (TSP), which is to find a shortest route for visiting n cities exactly once and finally returning to the origin, where the distance d_{ij} between any two cities i and j is known and symmetric. A TSP instance with six cities is illustrated in Fig. 1.3. TSP can be formulated as a 0–1 programming problem shown in Eqs. (1.7)–(1.11):

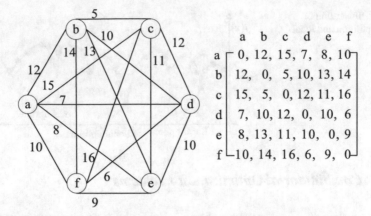

Fig. 1.3 An example of TSP. The right part of the figure gives the adjacent matrix, i.e., the matrix of the distances between pairs of cities

$$\min \sum_{i=1}^{n} d_{ij} x_{ij} \tag{1.7}$$

$$\text{s.t.} \sum_{i=1}^{n} x_{ij} = 1, \quad j = 1, 2, \dots, n \tag{1.8}$$

$$\sum_{j=1}^{n} x_{ij} = 1, \quad i = 1, 2, \dots, n \tag{1.9}$$

$$\sum_{i,j \in S} x_{ij} \leq |S| - 1, \quad 2 \leq |S| \leq n - 2; S \subset \{1, 2, \dots, n\} \tag{1.10}$$

$$x_{ij} \in \{0, 1\}, \quad i, j \in \{1, 2, \dots, n\}; i \neq j \tag{1.11}$$

As we can see, the model consists of $n(n-1)$ decision variables, where each $x_{ij} = 1$ denotes that the edge from i to j is included in the route and $x_{ij} = 0$ otherwise. Constraint (1.8) denotes that the route enters each city exactly once, (1.9) denotes that the route leaves each city exactly once, and (1.10) denotes that there is no cycle in any subset that consists of at least 2 but no more than $(n-2)$ cities.

The above formulation uses so many decision variables and constraints. Sometimes a problem can be formulated in different forms by using different domains and functions. For TSP, each solution is a permutation of the city set. Let perms(S) denote the set of all permutations of set S and $x[i]$ denote the ith elements of permutation vector $\mathbf{x}$; the TSP can be reformulated as the following simpler form:

$$\min \left(\sum_{i=1}^{n-1} d_{x[i]x[i+1]} \right) + d_{x[n]x[1]}$$

$$\text{s.t.} \ \ \mathbf{x} \in \text{perms}(\{1, 2, \dots, n\}) \tag{1.12}$$

Combinatorial optimization problems can also have local optima, though their physical representations may not be so obvious as in continuous ones.

Algorithms for optimization problems can be classified in different ways. According to whether they can guarantee to find the exact optimal solution (or, when the error is inevitable, a near-optimal solution within a predefined accuracy) to the problems, optimization algorithms can be divided into two classes: exact algorithms and heuristic algorithms.

1.3 Exact Optimization Algorithms

1.3.1 Gradient-Based Algorithms

For some optimization problems, the gradient information of objective functions can be very useful in problem-solving. In particular, for a one-dimensional real function optimization problem, if the function has first-order continuous derivative, then its extreme values are always at the points where the derivatives are zero. For example, for the problem shown in Eq. (1.2), we have $f'(x) = 12x^2 + 30x - 18 = 6(2x - 1)(x + 3)$, whose value is zero on the point $x = 0.5$ within the domain $-1 \leq x \leq 1$. It is easy to verify that $f(0.5) = -4.75$ is the minimum function value within the domain. In this way, the problem is solved directly.

Nevertheless, for many complex functions, it can be difficult or even impossible to solve their derivative equations. But we can still use some gradient-based algorithms to solve them. The classical Newton method is one of such algorithms. It uses the second-order Taylor expansion of the objective function to approximate the function and then iteratively searches for a point whose derivative is zero or very small as the optimal solution. Algorithm 1.1 presents the procedure for optimizing a one-dimensional objective function $f(x)$ (where ϵ is a small allowable error)

Algorithm 1.1: The Newton method for one-dimensional function optimization

---

1 Let $k = 0$ and select an initial $x^{(0)}$ in the search range;
2 **while** $|f'(x^{(k)})| > \epsilon$ **do**
3 $\quad$ $x^{(k+1)} \leftarrow x^{(k)} - f'(x^{(k)})/f''(x^{(k)})$;
4 $\quad$ $k \leftarrow k + 1$;
5 **return** $x^{(k)}$;

The process can be extended for multi-dimensional functions (where decision variable **x** is a vector) by simply replacing Line 4 of Algorithm 1.1 with Eq. (1.13):

$$x^{(k+1)} = x^{(k)} - \nabla^2 f(x^{(k)})^{-1} \nabla f(x^{(k)}) \tag{1.13}$$

where $\nabla^2 f(x^{(k)})^{-1}$ is the inverse of Hesse matrix $\nabla^2 f(x^{(k)})$.

Another widely used gradient-based algorithm is the steepest descent method, which always chooses the direction along which the function value descends fastest as the search direction. The procedure of this method is shown in Algorithm 1.2.

Algorithm 1.2: The steepest descent method

---

1 Let $k = 0$ and select an initial $x^{(0)}$ in the search range;
2 Calculate the search direction $d^{(k)} = -\nabla f(x^{(k)})$;
3 **while** $|d^{(k)}| > \epsilon$ **do**
4 $\quad$ Search along with $d^{(k)}$ to find a $\lambda^{(k)} \geq 0$ minimizing $f(x^{(k)} + \lambda^{(k)}d^{(k)})$;
5 $\quad$ $x^{(k+1)} \leftarrow x^{(k)} + \lambda^{(k)}d^{(k)}$;
6 $\quad$ $k \leftarrow k + 1$;
7 **return** $x^{(k)}$;

Gradient-based algorithms require that the objective function is differentiable (the Newton method also requires twice differentiable), which cannot be met by a variety of functions such as Eq. (1.3).

1.3.2   Linear Programming Algorithm

For a continuous optimization problem in real number space, if its objective and constraint functions are all linear, it is called a linear programming problem and otherwise a nonlinear programming problem. For example, the problem in Eq. (1.4) is linear, while the problem in Eq. (1.5) is nonlinear. A linear programming problem can be expressed by the standard model shown in Eq. (1.14) (where $\mathbf{c}$ is an n-dimensional vector, $\mathbf{b}$ is an m-dimensional vector, and $\mathbf{A}$ is an $m \times n$ matrix):

$$\min \ \mathbf{cx}$$
$$\text{s.t.} \ \ \mathbf{Ax} = \mathbf{b} \tag{1.14}$$
$$\mathbf{x} \geq 0$$

Linear programming problem (1.14) has a good feature that the set of all feasible solutions is a convex set. Taking Eq. (1.4) as an example, its constraints construct a convex polygon as shown in Fig. 1.4 (where the feasible region is indicated by the dashed area). It is easy to see that the minimum of the objective function $12x_1 + 16x_2$ can be gained when the contour line $12x_1 + 16x_2 = K$ crosses the vertex $(0.5, 0.375)$ of the polygon.

By extension, the optimal solution to any linear programming problem can always be obtained at some vertices of the feasible region. The most popular algorithm for linear programming is the simplex method [3] shown in Algorithm 1.3, which starts with one vertex and iteratively moves from one vertex to another better one until there

Fig. 1.4 Graphic solution of linear programming problem

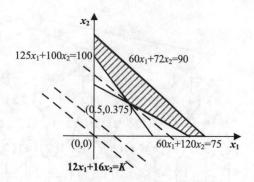

are no better vertices. However, the method cannot be applied to most problems which have nonlinear objective functions and/or constraints.

Algorithm 1.3: The Simplex method

---

1 Decompose matrix $\mathbf{A}$ into $[\mathbf{B}, \mathbf{N}]$, here $\mathbf{B}$ is an invertible basis matrix of order m;

2 **while** *true* **do**

3   Calculate a basic feasible solution $\mathbf{x} = \begin{bmatrix} \mathbf{B}^{-1}\mathbf{b} \\ 0 \end{bmatrix}$;

4   Let $k = \arg\max_{j \in R}\{\mathbf{c_B}\mathbf{B}^{-1}\mathbf{p}_j - \mathbf{c}_j\}$, where R is the set of subscripts of all non-basis variables, $\mathbf{P}_j$ is the jth column vector in $\mathbf{N}$, and c_j and x_j are the corresponding coefficient and independent variables, respectively;

5   **if** $\mathbf{c_B}\mathbf{B}^{-1}p_k - c_k \leq 0$ **then**

6   The current solution is optimal and **return**;

7   **if** *all the components of* $\mathbf{A}_k$ *is not positive* **then**

8   **return** as the problem has infinite optimal solutions.

9   Select a subscript r satisfying that b_r / A_{kr} is the minimum among all positive components of $\mathbf{A}_k$;

10   Exchange the rth and kth variables to a new basis $\mathbf{B}$;

1.3.3 Branch-and-Bound

For combinatorial optimization problems, the most direct method is to check all candidate solutions in the discrete solution space and finally select the best one. This is called exhaustive or brute force search. Branch-and-bound is a variant of exhaustive search, which improves the search efficiency by discarding each search direction if it can decide that the search along with the direction cannot lead to a solution better than the best objective function value f_c found so far.

Taking the TSP shown in Fig. 1.3 for example, suppose that the route 1–3–2–5–4–1 (whose length is 33) has been checked, when searching for a new solution, whenever the length of a part of the route (e.g., 1–2–4) is not less than 33, the route is

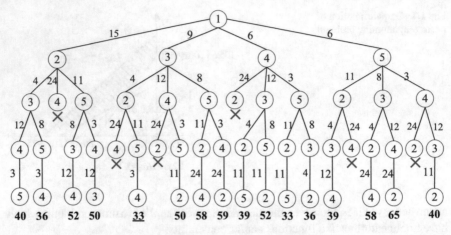

Fig. 1.5 Branch-and-bound process on the TSP instance in Fig. 1.3

discarded. Sometimes there are better methods for estimating the lower limit of the objective function value. For example, the minimum distance δ between the cities in Fig. 1.3 is 3, whenever the length of a part of the route (e.g., 1–4–2) is not less than 33, the route is discarded. Algorithm 1.4 shows the branch-and-bound procedure for TSP, and Fig. 1.5 illustrates the process to solve the instance shown in Fig. 1.3.

Algorithm 1.4: The branch-and-bound procedure for the TSP

---

1 Create an empty list L, select a departure city c_0 and add it into L_1, and set $f_c = +\infty$;
2 **while** L *is not empty* **do**
3   Take a partial path p from L, let k be the length of p;
4   **foreach** *city c in the remaining $(n - k)$ cities* **do**
5   Get a new path p' by adding c to p;
6          $f(p') = f(p) + d_{p[k],c}$;
7   **if** p' *has $(n - 1)$ cities* **then**
8              $f(p') = f(p) + d_{c,c_0}$;
9   Update $f_c = \min(f_c, f(p'))$;
10   **else if** $f(p') + \delta < f_c$ **then**
11   Add p' to L;

12 **return** *the shortest path found so far*;

Exhaustive method and its variants can only work well on small-size instances of combinatorial optimization problems. For example, suppose an exhaustive search method needs 1ms to solve a TSP instance of 5 cities by on a computer; i.e., the processor can calculate $5! = 120$ complete paths within 0.001s. Then, for an instance of 10 cities, the total number of complete paths is $10! = 3628800$, and the computational time will be more than 30.24s; for an instance of cities, the total of paths is 20! $\approx 2.4 \times 10^{18}$, and time will be more than 6.4×10^6 years. The brand-and-branch can partially improve the search efficiency, but cannot fundamentally address the issue of combinational explosion.

1.3.4 Dynamic Programming

Some combinatorial optimization problems can be decomposed into a set of subproblems, and the optimal solution to the original problem can be obtained by combining the optimal solutions to the subproblems. During the process of problem decomposition, some subproblems may occur many times. Once a subproblem is solved, we can save its solution in a memory; the next time the same subproblem occurs, we can reuse the saved solution instead of recomputing, and thereby save the computational time. This is the basic ideal of dynamic programming.

Taking the 0–1 knapsack problem shown in Eq. (1.6) for example, let $P(b, n)$ be the maximum value sum that can be attained with knapsack capacity b and item number n. We consider two cases according to size of the last item n:

- If $a_n \leq b$, we have two choices: Add the item to knapsack, then we get a value of c_n, and the maximum value of the remaining items is $P(b - a_n, n - 1)$; or do not add the item, then we get no value, and the maximum value of the remaining items is $P(b, n - 1)$.
- If $a_n > b$, we only need to consider the remaining $n - 1$ items, i.e., $P(b, n) = P(b, n - 1)$.

Thus we have

$$P(b, n) = \begin{cases} \max\left(P(b, n - 1), P(b - a_n, n - 1) + c_n\right), & a_n \leq b \\ P(b, n - 1), & a_n > b \end{cases} \qquad (1.15)$$

Starting with the simplest subproblems $P(0, j) = 0$ and $P(i, 0) = 0$, we can incrementally calculate $P(i, j)$ based on $P(i, j - 1)$ and $P(i - a_j, j - 1)$ for all $0 < i \leq b$ and $0 < j \leq n$, and finally get $P(b, n)$, the procedure of which is shown in Algorithm 1.5.

Algorithm 1.5: Dynamic programming for the 0–1 knapsack problem

---

1 Initialize a $(b + 1) \times (n + 1)$ matrix P;
2 **for** $i = 0$ *to* b **do** $P(i, 0) = 0$;
3 **for** $j = 0$ *to* n **do** $P(0, j) = 0$;
4 **for** $i = 1$ *to* b **do**
5   **for** $j = 1$ *to* n **do**
6   **if** $c_j \leq i$ **then** $P(i, j) = \max(P(i, j - 1), P(i - a_j, j - 1) + c_j)$
7   **else** $P(i, j) = P(i, j - 1)$

8 **return** $P(b, n)$;

In order to apply dynamic programming, a problem should have both the optimal substructure and overlapping subproblems, but most combinatorial optimization problems do not satisfy these requirements. TSP does not have the optimal substructure; i.e., the complete shortest tour usually does not consist of the shortest path between each pair of cities.

1.4 Heuristic Optimization Algorithms

As mentioned above, almost every exact optimization algorithm has one or more specific requirements for the problem to be solved. In comparison, heuristic algorithms typically do not have such requirements and can be applicable to a wide variety of problems. But most of them cannot guarantee the optimal solution; instead, their purpose is to find satisfactory solutions for complex and large-size problems within a reasonable time. Next we introduce some typical heuristic algorithms.

1.4.1 Genetic Algorithms

GA [9] is inspired by Darwinian evolutionary theory and Mendelian inheritance. According to the inheritance, each gene controls one or some features of it owning organism, and chromosomes consisting of genes largely determine the organism's fitness to the environment. According to the evolutionary theory, most features of an individual can be inherited by its offspring, and those features better fitted to the environment are more likely to be kept. The offspring can also adapt to the environment, while those individuals who are inadaptable to the environment will be eliminated gradually, which is known as "survival of the fittest."

In GA, each individual solution is analogous to a chromosome, each solution component is analogous to a gene, and the fitness of a solution is analogous to the chromosome's fitness to the environment. The fitter the solution, the more likely its genes to be transmitted to its offspring. By evolving a population of solutions from generation to generation, the algorithm will finally obtain an optimal or near-optimal solution. The main three operators in GA are selection, crossover, and mutation.

Selection Selection is to determine which individuals in the current population enter the next generation. According to "survival of the fittest," the fitter the individual, the higher the probability of being selected. But the selection procedure should be random in essence, and those solutions of low fitness still have some chances. The most popular selection method used in GA is called roulette wheel selection, in which the probability that each individual i in the population of N individuals is being selected is calculated by Eq. (1.16) (where f_i is the fitness of individual i):

$$p_i = \frac{f_i}{\sum_{j=1}^{N} f_j} \tag{1.16}$$

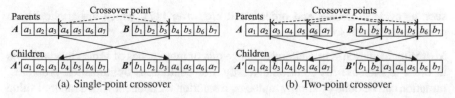

(a) Single-point crossover   (b) Two-point crossover

Fig. 1.6 Illustration of single-point and two-point crossover

Fig. 1.7 Illustration of
partially matched crossover
for TSP

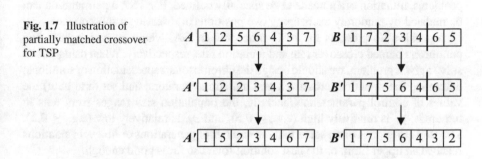

We can imagine that a circle (roulette wheel) has been divided into N sectors, and the size of each sector is proportional to the fitness of the corresponding solution; the selection is made by randomly rotating the wheel to get a sector.

For a maximization problem, the objective function values can be used as fitness; but for a minimization problem, the fitness should be inversely proportional to the objective values. Besides roulette wheel selection, there are other selection methods, such as ranking selection, tournament selection, and steady-state selection [8].

Crossover Offspring is generated by crossover (recombination). There are different crossover methods. A simple one is single-point crossover, where two parents exchange their genes from a random position to reproduce two offspring, as shown in Fig. 1.6a. This can be extended to multi-point crossover; e.g., Fig. 1.6b shows a two-point crossover operation.

Crossover depends on the chromosome representation, and inadequate crossover operations may produce illegal solutions. For TSP, if we use permutation-based representation, single-point crossover of two permutations often produces offspring which is not permutation; i.e., some cities are visited twice while some are not visited. In such a case, an effective crossover method is partially matched crossover (PMX), which first performs a basic two-point crossover operation and then inversely replaces the duplicated cities outside of the two crossover points with the corresponding cities in the parents. As shown in Fig. 1.7, two children A' and B' are generated by two-point crossover of parents A and B. For A', city 2 duplicates, and its second component is inversely replaced by city 5. The same procedure is applied to B'. Consequently, both A' and B' become legal permutations. There are also other crossover operators, such as cycle crossover and order-based crossover [17], which are applicable to permutation-based solutions.

Mutation Mutation is to randomly change one or more genes of some chromosomes in order to increase the solution diversity. For continuous optimization problems, mutation can be done by simply replacing a solution component with a random value in the feasible range. For example, for the problem shown in Eq. (1.5), component x can be replaced by a random value in [0,50]. But for combinatorial optimization problems, mutation often needs to be specially defined. For TSP, a permutation can be mutated by randomly exchanging two positions or reversing a subsequence.

Algorithm 1.6 presents the basic procedure of GA, where p_c and p_m are two parameters named crossover rate and mutation rate, respectively. When using GA to solve a given problem, we should design the chromosome representation of solutions; determine the selection, crossover, and mutation operators; and set (and tune) the values of control parameters. Generally, the population size ranges from tens to hundreds, p_c is relatively high (e.g., ≥ 0.5), and p_m is relatively low (e.g., ≤ 0.1). The stop criterion is often set as that the number of generations or fitness evaluations reached an upper limit, or the best solution found so far is good enough.

Algorithm 1.6: The basic GA

---

1 Initialize a population **P** of N solutions;
2 **while** *the stop criterion is not satisfied* **do**
3   Evaluate the fitness of the solutions, based on which select N solutions from **P** to a candidate population **P'**;
4   Empty **P**;
5   **while** $|\mathbf{P}| < N$ **do**
6   Randomly select two parents **x** and **x'** from **P'**;
7   With a probability of p_c, conduct crossover on **x** and **x'**;
8   With a probability of p_m, conduct mutation on **x** and **x'**;
9   Add **x** and **x'** to **P**;

10 **return** *the best solution found so far*;

There are also many variants of the basic GA, such as GA using double chromosomes (where each solution is represented by two chromosomes), self-adaptive GA (whose parameters are dynamically adjusted during the search), and GA combing with local search [22].

1.4.2 Simulated Annealing

Annealing is a physical process, where a metal material is cooled down from a high temperature and finally reaches a low energy, well-ordered final state. At each temperature T, the material is allowed to reach thermal equilibrium, characterized by a probability of being in a state s with energy E_s given by the Boltzmann distribution:

$$P_s(T) = C \cdot \exp(-E_s/T) \tag{1.17}$$

where C is a constant. Given two states 1 and 2 under the same temperature, we have

$$\frac{P_1(T)}{P_2(T)} = \frac{C \cdot \exp(-E_1/T)}{C \cdot \exp(-E_2/T)} = \exp\left(\frac{E_2 - E_1}{T}\right) \tag{1.18}$$

Equation (1.18) indicates that, if $E_1 > E_2$, the probability that the material stays in state 1 is less than that in state 2. When the temperature is the lowest, the probability that the material stays in the lowest energy state is 1.

SA [12] is a heuristic optimization algorithm simulating the physical annealing process. It regards each solution as a state and the solution's objective function value as the energy of its state. The algorithm starts from an initial solution at a high temperature $T^{(0)}$ and then iteratively lowers the temperature; at each temperature, it searches a number of neighboring solutions and probabilistically selects one as the new current state. Ultimately, the algorithm probably reaches a state with the lowest possible energy, i.e., the optimal solution to the problem. Assume that the problem is to minimize the objective function f, Algorithm 1.7 presents the SA algorithm, where $T^{(0)}$ and k_N are control parameters, and rand() produces a random number uniformly distributed in [0,1]. $T^{(0)}$ is usually set to a value far larger than the objective function values; temperature lowering is often done by setting $T - rT$ where r is a positive number less than 1, but more complex cooling methods can also be used.

Algorithm 1.7: The SA algorithm

---

1 Initialize an initial solution $\mathbf{x}^{(0)}$, let $x = \mathbf{x}^{(0)}$, $T = T^{(0)}$;
2 **while** *the stop criterion is not satisfied* **do**
3   **for** $\iota = 1$ *to* k_N **do**
4   Generate a neighboring solution $\mathbf{x}'$, calculate $\Delta f = f(\mathbf{x}) - f(\mathbf{x}')$;
5   **if** $\Delta f \leq 0$ *or* $\exp(-\Delta f/T) > $ rand() **then** set $\mathbf{x} = \mathbf{x}'$
6   Lower T;
7 **return** *the best solution found so far*;

SA is largely based on neighborhood search. For a continuous optimization problem, we can get neighboring solutions whose distances from $\mathbf{x}$ are less than a small threshold $\hat{d}$. For a combinatorial optimization problem, we can get neighbors by making some slight changes to $\mathbf{x}$. For example, for TSP, a neighbor can be obtained by exchanging two positions in the permutation.

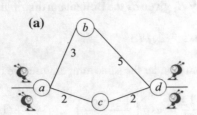 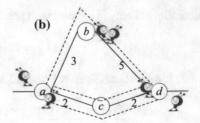

Fig. 1.8 Illustration of path selection by ants

1.4.3 Ant Colony Optimization

ACO [4] is a swarm intelligence method inspired by the foraging behavior of real ant colonies. When searching for food, ants leave pheromone trails on the path between food sources and their nests. But pheromone trails evaporate with time. The higher the quantity and quality of the food, the higher pheromone density, and the more ants will be attracted to the food source, which will further enhance the pheromone density on the path. On the contrary, the pheromone density on a path followed by few ants drops rapidly. Eventually, such a feedback loop often leads to the optimal path from the nest to the food source.

Figure 1.8 illustrates such a process. As shown in Fig. 1.8a, suppose that at beginning there are two ants leave node a for node d and two ants leave d for a. As the paths have not been explored by any ant, the probabilities of choosing path abd and choosing path acd are equal, and we assume that each path is selected by two ants. When the ants choosing the shorter path acd reach the end, the ants choosing the longer path abd are still on the way. As a consequence, any new ant at a or d will have a higher probability of choosing acd, and the pheromone density on acd will become higher than that on abd, as shown in Fig. 1.8b (where dotted lines indicate pheromone trails). Such a process can be extended for using a large number of ants to search the optimal path in a complex network.

ACO was first used to solve TSP, the procedure of which is shown in Algorithm 1.8, where α, β, and ρ are control parameters. Later it has been applied to solve many scheduling problems such as vehicle routing and job scheduling [1, 18]. When applying ACO to other optimization problems other than shortest paths, the main idea is to represent each solution to the problem in the form of "a path" and evaluate the solution as "the length of the path."

Algorithm 1.8: The ACO algorithm

---

1 Randomly place m ants on n cities, and let the initial pheromone on each path be zero;
2 **while** *the stop criterion is not satisfied* **do**
3   **for** $k = 1$ *to* m **do**
4   **while** *the number of cities visited by ant k is less than n* **do**
5   Let i be the current city at which ant k is located;
6   Among the set A_i of the adjacent cities of i, select the next city j to be visited by ant k, where the selection probability of each j is:

$$p_{ij}^k = \frac{(\tau_{ij})^\alpha (d_{ij})^{-\beta}}{\sum\limits_{l \in A_i} (\tau_{il})^\alpha (d_{il})^{-\beta}} \tag{1.19}$$

7   Get a complete path by making ant k return to the original city;
8   Calculate the pheromone increment Δ_k which is inversely proportional to the length of the path;
9   **foreach** *edge (i, j) on the path* **do**
10   Update the pheromone on (i, j) as:

$$\tau_{ij} = (1 - \rho)\tau_{ij} + \Delta_k \tag{1.20}$$

11   Clear the visited path of ant k;

12 **return** *the shortest tour found;*

1.4.4 Particle Swarm Optimization

PSO [10] is another swarm intelligence method inspired by social behavior of bird flocking. Similar to GA, PSO also evolves a population (swarm) of candidate solutions (called particles), but it does not use crossover and mutation operators. Instead, PSO moves the particles in the search space by making them learn from their own experiences as well as other good particles.

In an n-dimensional search space, every particle i has a position vector $x_i = (x_{i1}, x_{i2}, \ldots, x_{in})$ and a velocity vector $v_i = (v_{i1}, v_{i2}, \ldots, v_{in})$. During the search, the best-known position visited by particle i is recorded as **pbest**$_i$ and the best-known position of the entire swarm is denoted as **gbest**. At each iteration, each particle i updates its velocity and position at each dimension j as follows:

$$v_{ij} = wv_{ij} + c_1 \cdot \text{rand}() \cdot (pbest_{ij} - x_{ij}) + c_2 \cdot \text{rand}() \cdot (gbest_j - x_{ij}) \tag{1.21}$$

$$x_{ij} = x_{ij} + v_{ij} \tag{1.22}$$

where w is the inertia weight, and c_1 and c_2 are learning coefficients. The parameters can be constants or adjusted dynamically. Generally speaking, a larger w is better for the particles to explore new regions and a smaller w facilitates the particles to gather for exploitation. A recommended strategy is to make w decrease with iteration [6]:

$$w = w_{\max} - \frac{t}{t_{\max}}(w_{\max} - w_{\min}) \tag{1.23}$$

where t is the current iteration number, $t_{\max}$ is the maximum number of iterations, and $w_{\max}$ and $w_{\min}$ are lower and upper limits of w, respectively. Algorithm 1.9 presents the pseudocode of PSO.

Algorithm 1.9: The PSO algorithm

---

1 Initialize a swarm of solutions to the problem;
2 **while** *the stop termination criterion is not satisfied* **do**
3   Calculate the fitness the particles;
4   Update **pbest**s and **gbest**;
5   **for** *each particle i in the swarm* **do**
6   Update the particle's velocity according to Eq. (1.21);
7   Update the particle's position according to Eq. (1.22);
8   Update the control parameters;
9 **return** *the best solution found so far;*

The original PSO is for searching in continuous spaces. For combinatorial optimization problems, the operations of velocity and position updating need to be adapted to the problems, but the main idea is still the combination of learning from the personal best and learning from the global best. For example, when being applied to TSP, a particle can be updated by cloning subsequences from the personal best and the global best.

Like many other heuristic algorithms, PSO is also easy to be trapped in local optima. This is mainly because that, when **gbest** approaches a local optimum, the other particles will be attracted by **gbest** and thus move toward the local optimum. To address this issue, some invariants of PSO have been proposed, where each particle can learn from other exemplars, such as its best neighboring solution or a solution probabilistically selected, rather than the global best [11].

1.4.5 Differential Evolution

DE [23] is an efficient evolutionary algorithm for high-dimensional continuous optimization problems. Similar to GA, DE maintains a population of solutions and uses selection, crossover, and mutation to evolve the population. The key difference is that DE produces new solutions by combining existing ones according to their differences.

The basic DE algorithm is shown in Algorithm 1.10, where Eq. (1.24) calculates the difference between two randomly selected solutions and then add to the third one to create a mutation vector. This is called the DE/rand/1/bin scheme.

Algorithm 1.10: The DE algorithm

---

1 Initialize a population of N solutions;
2 **while** *the stop criterion is not satisfied* **do**
3   **forall the** $\mathbf{x}_i$ *in the population* **do**
4   Create a mutation vector $\mathbf{v}_i$ as:

$$\mathbf{v}_i = \mathbf{x}_{r_1} + F \cdot (\mathbf{x}_{r_2} - \mathbf{x}_{r_3}) \tag{1.24}$$

        where F is a scale factor in [0,1], and r_1, r_2 and r_3 are three distinct random indices of the population;
5   Create a trial vector $\mathbf{u}_i$ by setting each dimension j from either $\mathbf{x}_i$ or $\mathbf{v}_i$:

$$u_{ij} = \begin{cases} v_{ij}, & (\text{rand}() \leq c_r) \vee (j = r_i) \\ x_{ij}, & \text{otherwise} \end{cases} \tag{1.25}$$

        where c_r is the crossover rate, and r_i is a random integer in $[0, N]$ to ensure that there is at least one dimension is from $\mathbf{v}_i$;
6   Select the fitter one among $\mathbf{x}_i$ and $\mathbf{u}_i$ to replace $\mathbf{x}_i$ in the population;

7 **return** *the best solution found so far;*

Nevertheless, DE can use other mutation schemes as follows (where $\mathbf{x}_{\text{best}}$ denotes the best solution found so far):

- DE/best/1/bin

$$\mathbf{v}_i = \mathbf{x}_{\text{best}} + F \cdot (\mathbf{x}_{r_1} - \mathbf{x}_{r_2}) \tag{1.26}$$

- DE/rand/2/bin

$$\mathbf{v}_i = \mathbf{x}_{r_1} + F \cdot (\mathbf{x}_{r_2} - \mathbf{x}_{r_3}) + F \cdot (\mathbf{x}_{r_4} - \mathbf{x}_{r_5}) \tag{1.27}$$

- DE/best/2/bin

$$\mathbf{v}_i = \mathbf{x}_{\text{best}} + F \cdot (\mathbf{x}_{r_1} - \mathbf{x}_{r_2}) + F \cdot (\mathbf{x}_{r_3} - \mathbf{x}_{r_4}) \tag{1.28}$$

- DE/rand-to-best/2/bin

$$\mathbf{v}_i = \mathbf{x}_i + F \cdot (\mathbf{x}_{\text{best}} - \mathbf{x}_i) + F \cdot (\mathbf{x}_{r_1} - \mathbf{x}_{r_2}) \tag{1.29}$$

When applying DE for combinatorial optimization problems, we need to redefine the "difference" between any two solutions, so that the mutation can work on the discrete solution space.

1.4.6 Harmony Search

Different from those bio-inspired algorithms, HS [7] is an optimization algorithm inspired by the improvisation process of musicians. When composing a harmony, a musician often plays a note in one of the three means: taking a note from his/her memory, tweaking a note from his/her memory, or randomly selecting a note from the total playable range. The musician continually searches for a better state of harmony until reaching a perfect state.

In HS, each solution is analogous to a harmony, a solution component is analogous to a note, and the solution fitness is analogous to the quality of the harmony. HS also maintains a population (memory) of solutions (harmonies), but at each iteration, it produces only one new harmony based on the three means mentioned above. The basic HS algorithm is shown in Algorithm 1.11, where the parameter *HMCR* (harmony memory considering rate) controls the probability that a component comes from the memory, *PAR* (pitch adjusting rate) controls the probability that a component will be tweaked, and *bw* (bandwidth) controls the range of tweaking.

Algorithm 1.11: The HS algorithm

---

1 Initialize a population of N solutions;
2 **while** *the stop criterion is not satisfied* **do**
3   Generate a new empty harmony x_{new};
4   **forall the** *dimension j of x_{new}* **do**
5   **if** rand$()$ > *HMCR* **then**
6   Set $x_{new,j}$ as a random value in the range;
7   **else**
8   Select a random solution $\mathbf{x}_r$ from the memory and set $x_{new,j} = x_{r,j}$;
9   **if** rand$()$ $\leq$ *PAR* **then**
10   Tweak the component as:

$$x_{new,j} = x_{new,j} + \text{rand}(-1, 1) \cdot bw \qquad (1.30)$$

11   **if** *x_{new} is better than the worst harmony in the memory* **then**
12   replace the worst harmony with x_{new};

13 **return** *the best solution found;*

For real-valued function optimization problems, *HMCR* is typically set to a value between 0.7 and 0.9, *PAR* is set to a value between 0.1 and 0.5, and *bw* is set according to the search range. A variant of HS decreases the values of *PAR* and *bw* with iteration, so as to facilitate global search in early stages and enhance local search in later stages [14].

When applying HS to a combinatorial optimization problem, we can use specific local search (such as neighborhood search) to replace the pitch adjusting operator (1.30), while the other operators as well as the algorithm framework do not need to be much changed.

Fig. 1.9 Illustration of firework explosion

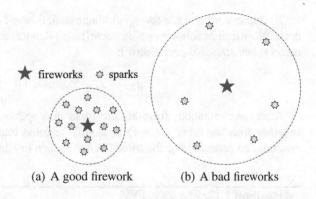

★ fireworks ✿ sparks

(a) A good firework (b) A bad fireworks

1.4.7 Fireworks Algorithm

FWA [24] is an optimization algorithm simulating the explosion of fireworks in the night sky, where the search space is analogous to the sky, and each solution is analogous to a firework or spark. FWA first chooses a set of random positions and sets off a firework in each position to generate sparks. The core idea of FWA is that better fireworks generate more sparks in smaller ranges, and worse fireworks generate less sparks in larger ranges, shown in Fig. 1.9. In this process, the quality of the population is gradually improved and promising results will be yielded.

Assume that the problem is to minimize the objective function f, FWA calculates the explosion amplitude A_i of and the number s_i of sparks generated by each firework $\mathbf{x}_i$ as follows:

$$A_i = \widehat{A} \frac{f(\mathbf{x}_i) - f_{\min} + \epsilon}{\sum_{j=1}^{N} (f(\mathbf{x}_j) - f_{\min}) + \epsilon} \tag{1.31}$$

$$s_i = M_e \frac{f_{\max} - f(\mathbf{x}_i) + \epsilon}{\sum_{j=1}^{N} (f_{\max} - f(\mathbf{x}_i)) + \epsilon} \tag{1.32}$$

where M_e and $\widehat{A}$ are two control parameters, $f_{\max}$ and $f_{\min}$ are the maximum and minimum objective values in the population, and ϵ is a very small value to avoid division-by-zero. In addition, To avoid overwhelming effects of some splendid fireworks, s_i is further limited by a lower $s_{\min}$ and upper $s_{\max}$.

To generate a new spark $\mathbf{x}_j$, the algorithm randomly chooses about half of dimensions of $\mathbf{x}_i$, and at each dimension d set:

$$x_{j,d} = x_{i,d} + A_i \cdot \text{rand}(-1, 1) \tag{1.33}$$

To improve diversity, a few sparks are mutated based on Gaussian distribution at each dimension as follows (where $\text{norm}(\mu, \sigma)$ denotes a Gaussian distribution with mean μ and standard deviation σ):

$$x_{j,d} = x_{i,d} \cdot \text{norm}(1, 1) \tag{1.34}$$

After each iteration, from all fireworks and sparks, the current best solution together with the other $(N - 1)$ randomly selected ones enter into the next generation. The procedure of the basic FWA is shown in Algorithm 1.12.

Algorithm 1.12: The basic FWA

---

1 Initialize a population of N fireworks;
2 **while** *the stop criterion is not satisfied* **do**
3   Evaluate the fireworks' fitness, based which calculate their A_i and s_i based on Eqs. (1.31) and (1.32);
4   **forall the** $\mathbf{x}_i$ *in the population* **do**
5   **for** $k = 1$ *to* s_i **do**
6   Generate a spark $\mathbf{x}_j$ based on Eq. (1.33);
7   With a mutation probability, mutate $\mathbf{x}_j$ based on Eq. (1.34);
8   Select best solution together with the other $(N - 1)$ for the next generation;

9 **return** *the best solution found;*

To overcome some shortcomings of the basic FWA, Zheng et al. [28] propose an enhanced FWA (EFWA) that improves FWA by setting different bounds for each dimension of the explosion amplitude and enabling learning from the best solution in Gaussian mutation as shown in (1.35). It becomes the current norm of FWA.

$$x_{j,d} = x_{i,d} \cdot (x_{\text{best},d} - x_{i,d}) \cdot \text{norm}(0, 1) \tag{1.35}$$

1.4.8 Water Wave Optimization

WWO [29] is a recent optimization algorithm inspired by shallow water wave models [5]. In WWO, each solution $\mathbf{x}$ is analogous to a wave. The higher the energy (fitness), the smaller the wavelength $\lambda_{\mathbf{x}}$, and thus the smaller the range the wave propagates, as shown in Fig. 1.10. In a continuous search space, the propagation operation is done by shifting each dimension d of the solution as follows ($L(d)$ denotes the length of the dth dimension of the search space):

$$x'(d) = x(d) + \lambda_{\mathbf{x}} \cdot \text{rand}(-1, 1) \cdot L(d) \tag{1.36}$$

Fig. 1.10 Illustration of wave propagation in shallow water

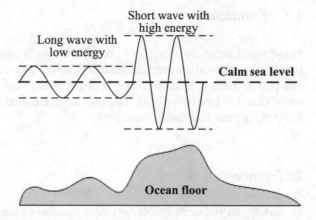

All wavelengths are initially set to 0.5 and then updated after each generation as follows:

$$\lambda_{\mathbf{x}} = \lambda_{\mathbf{x}} \cdot \alpha^{-(f(\mathbf{x})-f_{\min}+\epsilon)/(f_{\max}-f_{\min}+\epsilon)} \tag{1.37}$$

where α is the wavelength reduction coefficient set to 1.0026, and ϵ is a very small number to avoid division-by-zero.

Even using only the propagation operator, WWO can efficiently optimize many function optimization problems. However, to make it competitive with other state of the arts, WWO has two other operators:

- Refraction. If a wave has not been improved after several propagations, it loses its momentum and is removed. A new wave is generated at a random position between the old wave and the known best solution.
- Breaking. Whenever a new best wave $\mathbf{x}^*$ is found, it breaks into several solitary waves, each of which moves a small distance from $\mathbf{x}^*$ at a random direction.

The breaking is essentially a local search operator. The refraction operator, however, is discarded in an improved version of WWO [30] which gradually reduces the population size by removing the current worst solution.

We have compared WWO with a number of recent heuristic algorithms including BBO, invasive weed optimization [15], GSA [19], bat algorithm [26], and hunting search [16] on a benchmark set which consists of unimodal, multimodal, hybrid, and composition functions [13]. The results show that WWO exhibits very competitive performance.

Equation (1.37) is for continuous optimization, but we can easily adapt the propagation operator to combinatorial optimization, the basic idea of which is still to make low fitness solutions explore in remote areas while making high fitness solutions exploit in small neighboring regions, so as to achieve a good dynamic balance between diversification and intensification. In the last three years, WWO has aroused great research interest and has been adapted to solve many combinatorial optimization problems [20, 25, 27].

1.5 Summary

In our world, there are a variety of optimization problems. Some simple problems or problems with some specific features can be solved by exact algorithms. But for most complex problems, heuristic algorithms are widely used. This chapter introduces some classical heuristic algorithms, and since the next chapter, we will focus on BBO algorithms and their applications.

References

1. Bell JE, McMullen PR (2004) Ant colony optimization techniques for the vehicle routing problem. Adv Eng Inf 18:41–48. https://doi.org/10.1016/j.aei.2004.07.001
2. D, K (2005) An idea based on honey bee swarm for numerical optimization. Technical report, Erciyes University
3. Dantzig BG (1998) Linear programming and extensions. Princeton University Press, Princeton
4. Dorigo M, Maniezzo V, Colorni A (1996) Ant system: optimization by a colony of cooperating agents. IEEE Trans Syst Man Cybern Part B 26:29–41. https://doi.org/10.1109/3477.484436
5. Durran D (1999) Numerical methods for wave equations in geophysical fluid dynamics. Springer, Berlin
6. Eberhart R, Yuhui S (2001) Tracking and optimizing dynamic systems with particle swarms. Proc IEEE Congr Evol Comput 1:94–100. https://doi.org/10.1109/CEC.2001.934376
7. Geem ZW, Kim JH, Loganathan G (2001) A new heuristic optimization algorithm: harmony search. Simulation 76:60–68. https://doi.org/10.1177/003754970107600201
8. Goldberg DE, Deb K (1991) A comparative analysis of selection schemes used in genetic algorithms. Found Genetic Algorithms 1:69–93. https://doi.org/10.1016/B978-0-08-050684-5.50008-2
9. Henry HJ (1975) Adaptation in natural and artificial systems: an introductory analysis with applications to biology, control, and artificial intelligence. University of Michigan Press, Ann Arbor
10. Kennedy J, Eberhart R (1995) Particle swarm optimization. IEEE Int Conf Neural Netw 4:1942–1948. https://doi.org/10.1109/ICNN.1995.488968
11. Kennedy J, Mendes R (2006) Neighborhood topologies in fully informed and best-of-neighborhood particle swarms. IEEE Trans Syst Man Cybern C 36:515–519. https://doi.org/10.1109/TSMCC.2006.875410
12. Kirkpatrick S, Jr DG, Vecchi MP (1983) Optimization by simulated annealing. Science 220:671–680. https://doi.org/10.1126/science.220.4598.671
13. Liang JJ, Qu BY, Suganthan PN (2014) Problem definitions and evaluation criteria for the CEC 2014 special session and competition on single objective real-parameter numerical optimization. Technical report, Computational Intelligence Laboratory, Zhengzhou University
14. Mahdavi M, Fesanghary M, Damangir E (2007) An improved harmony search algorithm for solving optimization problems. Appl Math Comput 188:1567–1579. https://doi.org/10.1016/j.amc.2006.11.033
15. Mehrabian A, Lucas C (2006) A novel numerical optimization algorithm inspired from weed colonization. Ecol Inf 1:355–366. https://doi.org/10.1016/j.ecoinf.2006.07.003
16. Oftadeh R, Mahjoob M, Shariatpanahi M (2010) A novel meta-heuristic optimization algorithm inspired by group hunting of animals: hunting search. Comput Math Appl 60:2087–2098. https://doi.org/10.1016/j.camwa.2010.07.049
17. Potvin JY (1996) Genetic algorithms for the traveling salesman problem. Ann Op Res 63:337–370

18. Rajendran C, Ziegler H (2004) Ant-colony algorithms for permutation flowshop scheduling to minimize makespan/total flowtime of jobs. Eur J Op Res 155:426–438. https://doi.org/10.1016/S0377-2217(02)00908-6
19. Rashedi E, Nezamabadi-pour H, Saryazdi S (2009) GSA: A gravitational search algorithm. Inf Sci 179:2232–2248. https://doi.org/10.1016/j.ins.2009.03.004
20. Shao Z, Pi D, Shao W (2017) A novel discrete water wave optimization algorithm for blocking flow-shop scheduling problem with sequence-dependent setup times. Swarm Evol Comput. https://doi.org/10.1016/j.swevo.2017.12.005.(onlinefirst)
21. Simon D (2008) Biogeography-based optimization. IEEE Trans Evol Comput 12:702–713. https://doi.org/10.1109/TEVC.2008.919004
22. Srinivas M, Patnaik L (1994) Genetic algorithms - a survey. Computer 27:17–26. https://doi.org/10.1109/2.294849
23. Storn R, Price K (1997) Differential evolution - a simple and efficient heuristic for global optimization over continuous spaces. J Global Optim 11:341–359. https://doi.org/10.1023/A:1008202821328
24. Tan Y, Zhu Y (2010) Fireworks algorithm for optimization. In: Advances in swarm intelligence, Lecture Notes in Computer Science, vol 6145. Springer, pp 355–364. https://doi.org/10.1007/978-3-642-13495-1_44
25. Wu XB, Liao J, Wang ZC (2015) Water wave optimization for the traveling salesman problem. In: Huang DS, Bevilacqua V, Premaratne P (eds) Intelligent computing theories and methodologies. Springer, Cham, pp 137–146 (2015). https://doi.org/10.1007/978-3-319-22180-9_14
26. Yang XS (2010) A new metaheuristic bat-inspired algorithm. In: Nature inspired cooperative strategies for optimization, Studies in computational intelligence, vol 284. pp 65–74. https://doi.org/10.1007/978-3-642-12538-6_6
27. Zhao F, Liu H, Zhang Y, Ma W, Zhang C (2018) A discrete water wave optimization algorithm for no-wait flow shop scheduling problem. Expert Syst Appl 91:347–363. https://doi.org/10.1016/j.eswa.2017.09.028
28. Zheng S, Janecek A, Tan Y (2013) Enhanced fireworks algorithm. In: Proceedings of IEEE Congress on Evolutionary Computation. pp 2069–2077. https://doi.org/10.1109/CEC.2013.6557813
29. Zheng YJ (2015) Water wave optimization: a new nature-inspired metaheuristic. Comput Op Res 55:1–11. https://doi.org/10.1016/j.cor.2014.10.008
30. Zheng YJ, Zhang B (2015) A simplified water wave optimization algorithm. In: Proceedings of IEEE Congress on Evolutionary Computation. pp 807–813. https://doi.org/10.1109/CEC.2015.7256974

Chapter 2
Biogeography-Based Optimization

Abstract Biogeography is a discipline of the distribution, migration, and extinction of biological populations in habitats. Biogeography-based optimization (BBO) is a heuristic inspired by biogeography for optimization problems, where each solution is analogous to a habitat with an immigration rate and an emigration rate. BBO evolves a population of solutions by continuously migrating features probably from good solutions to poor solutions. This chapter introduces the basic BBO and its recent advances for constrained optimization, multi-objective optimization, and combinatorial optimization.

2.1 Introduction

Geographical distribution of species is a very attractive natural phenomenon, behind which we can find interesting optimization mechanisms: The origination of a species has a lot of randomness. However, a large number of species, after millions of years of migration, formed a complex, diverse, and magnificent ecological geography in nature. Inspired by this, in 2008 Simon [31] proposed a new metaheuristic algorithm called biogeography-based optimization (BBO), which borrows ideas from island biogeographical evolution to solve optimization problems. Although its scheme is simple, BBO has shown performance advantages over many other heuristics such as GA and PSO on a variety of optimization problems and has aroused widespread interest in the community of evolutionary optimization.

2.2 Background of Biogeography

The science of biogeography is an interdiscipline of life science and earth science that studies the distribution, migration, and extinction of biological populations in their habitats. As early as the nineteenth century, the works of Wallace [39] and

© Springer Nature Singapore Pte Ltd. and Science Press, Beijing 2019
Y. Zheng et al., *Biogeography-Based Optimization: Algorithms and Applications*,
https://doi.org/10.1007/978-981-13-2586-1_2

27

Darwin [8] had laid the foundation of biogeography. However, until the first half of the twentieth century, the research of biogeography mainly focused on the history of biogeographical distribution of terrestrial species. In the 1960s, MacArthur and Wilson began working together on mathematical models of the biogeographic distribution in islands, and in 1967, they published the classic work *The Theory of Island Biogeography* [23]. Since then, biogeography has become a mainstream research [16].

In nature, biological species are distributed in a series of geographical areas with clear boundaries, and such areas are called "habitats." Habitats that are well suited as residences for biological species are said to have a high habitat suitability index (HSI) [40]. There are many factors, including rainfall, vegetation diversity, geographical diversity, area size, temperature, that can have effects on HSI. Such factors are called suitability index variables (SIVs).

In general, high HSI habitats have a large number of species, while low HSI habitats have a small number of species. For a high HSI habitat, few outside species can move in because the habitat is already nearly saturated with species, and thus, the immigration rate of the habitat is low. Moreover, due to reasons such as intense competition among existing species, many species are likely to move to the nearby habitats, and thus, the emigration rate of the habitat is high. Consequently, the overall ecological status of high HSI habitats is relatively stable.

In contrast, low HSI habitats have low emigration rates and high immigration rates. Their HSI might be improved due to the migration of new species. However, if the HSI of a habitat remains at a low level, those immigrated species may be at risk of extinction. The extinction of some existing species will also open more opportunities for many other new species to immigrate. Consequently, the ecological status of low HSI habitats changes more dynamically than high HSI ones.

Figure 2.1 presents a simple mathematical model of species diversity in a habitat [23], where λ denotes the immigration rate and μ denotes the emigration rate, both of which are functions of s, the number of species in the habitat. I and E denote the maximum possible immigration rate and emigration rate of the habitat, respectively.

First, let us observe the immigration curve. When the number of species in the habitat is zero, the immigration rate $\lambda = I$. With the increase of the number of species, the habitat becomes more and more crowded, and the species that can successfully immigrate to and survive in the habitat become fewer and fewer, so the immigration rate λ decreases. When the number of species reaches the maximum possible number s_{max} that the habitat can support, λ decreases to zero.

Then, let us consider the emigration curve. When there are no species in the habitat, the emigration rate μ is certainly zero. With the increase of the number of species, the habitat becomes more and more crowded, and more species will leave the habitat to explore other new possible residences, so the emigration rate μ increases. When the number of species reaches s_{max}, we have $\mu = E$.

At the intersection of the immigration curve and the emigration curve, the immigration rate and the emigration rate are equal, and the corresponding number of species is s_0, which is called the equilibrium number of species. However, some unusual event may cause the number s to deviate from the equilibrium state. The

Fig. 2.1 A simple migration
model of a habitat

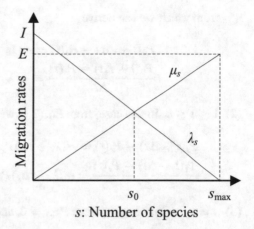

causes of positive deviations (growth) include a spurt of influx of species (such as
those dispossessed from neighboring habitats), a burst of speciation (like the Cambrian explosion). The causes of negative deviations (reduction) include diseases,
natural disasters, the entry of predators. After a great disturbance, it often takes a
long time for the number of species to recover to the equilibrium.

The immigration and emigration curves shown in Fig. 2.1 represent a simple linear
model. Although in many cases we need more complicated nonlinear models, their
basic features are the same or similar. Next, we perform a deeper mathematical
analysis based on the linear model.

Let P_s denote the probability that a habitat contains exactly s species. The change
of P_s from time t to time $(t + \Delta t)$ can be described as:

$$P_s(t + \Delta t) = P_s(t)(1 - \lambda_s \Delta t - \mu_s \Delta t) + P_{s-1}\lambda_{s-1}\Delta t + P_{s+1}\mu_{s+1}\Delta t \qquad (2.1)$$

where λ_s and μ_s are the immigration and emigration rates when there are s species.

Assuming that Δt is small enough so that we only need to consider the immigration
or emigration of at most one species. Under this assumption, Eq. (2.1) indicates that,
in order to accommodate s species at time $(t + \Delta t)$, the habitat must satisfy one of
the following conditions:

(1) There were s species at time t, and there is no immigration or emigration occurred
    during $[t, t + \Delta t]$.
(2) There were $(s - 1)$ species at time t, and only one species immigrated.
(3) There were $(s + 1)$ species at time t, and only one species emigrated.

Let $n = s_{max}$ for notational simplicity. According to different values of s, we can
analyze the change of P_s in the following three cases:

(1) $s = 0$. In this case, we have $P_{s-1} = 0$, and Eq. (2.1) becomes

$$P_0(t + \Delta t) = P_0(t)(1 - \lambda_0 \Delta t - \mu_0 \Delta t) + \mu_1 P_1 \Delta t$$

from which we can derive

$$P_0(t + \Delta t) - P_0(t) = -(\lambda_0 + \mu_0)P_0(t)\Delta t + \mu_1 P_1 \Delta t$$
$$\frac{P_0(t + \Delta t) - P_0(t)}{\Delta t} = -(\lambda_0 + \mu_0)P_0(t) + \mu_1 P_1$$

(2) $1 \leq s \leq n$. In this case, from Eq. (2.1) we can derive

$$P_s(t + \Delta t) - P_s(t) = -(\lambda_s + \mu_s)P_s \Delta t + \lambda_{s-1} P_{s-1} \Delta t + \mu_{s+1} P_{s+1} \Delta t$$
$$\frac{P_s(t + \Delta t) - P_s(t)}{\Delta t} = -(\lambda_s + \mu_s)P_s + \lambda_{s-1} P_{s-1} + \mu_{s+1} P_{s+1}$$

(3) $s = n$. In this case, we have $P_{s+1} = 0$, and Eq. (2.1) becomes

$$P_n(t + \Delta t) = P_n(t)(1 - \lambda_n \Delta t - \mu_n \Delta t) + \lambda_{n-1} P_{n-1} \Delta t$$

from which we can derive

$$P_n(t + \Delta t) - P_n(t) = -(\lambda_n + \mu_n)P_n \Delta t + \lambda_{n-1} P_{n-1} \Delta t$$
$$\frac{P_n(t + \Delta t) - P_n(t)}{\Delta t} = -(\lambda_n + \mu_n)P_n + \lambda_{n-1} P_{n-1}$$

Let $\dot{P}_s = \lim_{\Delta t \to 0} \frac{P_s(t+\Delta t) - P_s(t)}{\Delta t}$; by combining the above three cases, we have

$$\dot{P}_s = \begin{cases} -(\lambda_s + \mu_s)P_s + \mu_{s+1} P_{s+1}, & s = 0 \\ -(\lambda_s + \mu_s)P_s + \lambda_{s-1} P_{s-1} + \mu_{s+1} P_{s+1}, & 1 \leq s \leq n \\ -(\lambda_s + \mu_s)P_s + \lambda_{s-1} P_{s-1}, & s = n \end{cases} \quad (2.2)$$

Let $\mathbf{P} = [P_0 \dots P_n]^T$; Eq. (2.2) can be represented in the matrix form as

$$\dot{\mathbf{P}} = \mathbf{A}\mathbf{P} \quad (2.3)$$

where $\mathbf{A}$ is defined as

$$\mathbf{A} = \begin{bmatrix} -(\lambda_0 + \mu_0) & \mu_1 & 0 & \cdots & & 0 \\ \lambda_0 & -(\lambda_1 + \mu_1) & \mu_2 & \ddots & & \vdots \\ \vdots & \ddots & \ddots & \ddots & & \vdots \\ \vdots & & \ddots & \lambda_{n-2} & -(\lambda_{n-1} + \mu_{n-1}) & \mu_n \\ 0 & & \cdots & 0 & \lambda_{n-1} & -(\lambda_n + \mu_n) \end{bmatrix}$$

For the curves shown in Fig. 2.1, the immigration and emigration rates are as follows:

$$\mu_s = \frac{s}{n} E \tag{2.4}$$

$$\lambda_s = I(1 - \frac{s}{n}) \tag{2.5}$$

Consider a special case where $E = I = 1$, we have $\lambda_s + \mu_s = E$, and matrix A becomes

$$\mathbf{A} = E \begin{bmatrix} -1 & \frac{1}{n} & 0 & \cdots & 0 \\ \frac{n}{n} & -1 & \frac{2}{n} & \ddots & \vdots \\ \vdots & \ddots & \ddots & \ddots & \vdots \\ \vdots & \ddots & \frac{2}{n} & -1 & \frac{n}{n} \\ 0 & \cdots & 0 & \frac{1}{n} & -1 \end{bmatrix} \tag{2.6}$$

$$= E\mathbf{A}'$$

Obviously, zero is an eigenvalue of $\mathbf{A}'$, and the corresponding eigenvector is

$$\mathbf{v} = [v_1, \ldots, v_{n+1}]^T \tag{2.7}$$

where

$$v_i = \begin{cases} \frac{n!}{(n-1-i)!(i-1)!}, & i \le (n+1)/2 \\ v_{n+2-i}, & i > (n+1)/2 \end{cases} \tag{2.8}$$

For example, for $n = 4$ we have $\mathbf{v} = [1, 1, 4, 4, 6]^T$; for $n = 5$, we have $\mathbf{v} = [1, 1, 5, 5, 10]^T$.

Lemma 2.1 *The eigenvalue of $\mathbf{A}'$ is given by*

$$\{0, \frac{-2}{n}, \frac{-4}{n}, \ldots, -2\}$$

Theorem 2.1 *The steady-state value for the probability of the number of species is given by*

$$\mathbf{P}(\infty) = \frac{\mathbf{v}}{\sum_{i=1}^{n+1} v_i} \tag{2.9}$$

Interested readers can refer to Refs. [20, 31] for detailed proofs of Lemma 2.1 and Theorem 2.1.

2.3   The Basic Biogeography-Based Optimization
Algorithm

Inspired by the theory of biogeography, Simon [31] proposed the BBO algorithm
for optimization problems. As many other heuristics, BBO solves a problem by
continually evolving a population of solutions to the problem. The distinct features
of BBO include that it regards each solution as a habitat, the solution fitness as the
HSI of the habitat, and each solution component as an SIV. The evolution of solutions
in BBO is done by mimicking the migration and mutation process in biogeography.
Thus, *migration* and *mutation* are the two main operators of BBO.

2.3.1   The Migration Operator

In BBO, each solution H_i is analogous to a habitat and is assigned with an immigration
rate λ_i and an emigration rate μ_i. The higher the solution fitness, the higher the μ_i
and the lower the λ_i. On the contrary, the lower the fitness, the lower μ_i and the
higher the λ_i.

The main role of the migration operation is to share information among solutions:
Good solutions tend to share their information with others, while poor solutions tend
to receive information from others. At each iteration of the algorithm, BBO checks
the components of each solution H_i in the population one by one, uses a probability
proportional to λ_i to determine whether a component will be immigrated, and, if so,
chooses an emigrating solution H_j with a probability proportional to the emigration
rate μ_j, and replaces the current component of H_i with the corresponding component
of H_j. After all the components of H_i have been checked, a new solution $H_{i'}$ is
generated, and the better one between H_i and $H_{i'}$ remains in the population.

The procedure of migration can be described by Algorithm 2.1, where D is the
dimension of the problem, i.e., the length of the solution vector. Line 3 of Algorithm 2.1 means that the probability of each solution H_j being the emigrating solution is proportional to its emigration rate μ_j. This can be done by the roulette wheel
selection procedure of GA described in Sect. 1.3.

Algorithm 2.1: The migration operator in BBO

---
1 **for** $d = 1$ *to* D **do**
2   **if** rand() $< \lambda_i$ **then**
3   Select another H_j from the population with a probability $\propto \mu_j$;
4   Set $H_i(d) = H_j(d)$;

It is clear that the migration operation is also a random operation: Each component
in an immigrating solution may be, but is not necessarily to be replaced by the

corresponding component of the emigrating solution randomly selected. Although the probability is low, the components of good solutions may also be immigrated and the components of poor solutions may also be emigrated.

2.3.2 The Mutation Operator

In nature, some cataclysmic events can dramatically change a habitat and thereby cause significant effects on the species in the habitat. The BBO algorithm models such events as mutations. Equation (2.2) describes the probability of the number of species in a habitat. When the number of species is too large or too small, the probability is relatively low; when the number is medium (close to the equilibrium point), the probability is high. Taking $n = 5$ as an example, as indicated by Theorem 2.1, the steady-state values of the species are

$$\mathbf{P}(\infty) = \left[\frac{1}{32}, \frac{5}{32}, \frac{10}{32}, \frac{10}{32}, \frac{5}{32}, \frac{1}{32} \right]^{\mathrm{T}}$$

The above values are centrally symmetrical, as shown in Fig. 2.2.

Accordingly, BBO associates each solution H_i in the population with a species count probability P_i: Both high HSI habitats and low HSI habitats have low probabilities, while medium HSI habitats have high probabilities. The mutation rate π_i of each H_i is inversely proportional to the species count probability:

$$\pi_i = \pi_{\max} \left(\frac{1 - P_i}{P_{\max}} \right) \tag{2.10}$$

where $\pi_{\max}$ is a control parameter representing the maximum mutation rate.

Fig. 2.2 Distribution of species count probabilities

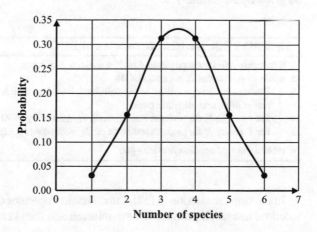

At each iteration, BBO checks the components of each solution H_i one by one, uses a probability proportional to π_i to determine whether a component will be mutated, and, if so, randomly sets the component to an allowed value. Suppose that the problem is a continuous optimization problem and the search range of the dth dimension is $[l_d, u_d]$, the procedure of BBO mutation can be described by Algorithm 2.2.

Algorithm 2.2: The mutation operator in BBO

---
1 **for** $d = 1$ *to* D **do**
2   **if** rand() $< \pi_i$ **then**
3   Set $H_i(d) = l_d + \text{rand}() * (u_d - l_d)$;

The main role of the migration operation is to improve the solution diversity. Low HSI habitats are likely to be mutated and thus have more opportunities to improve their fitness. High HSI habitats are also likely to be mutated, in order to weaken their dominance in the population and thus avoid premature convergence. In practice, we can use some elite strategies to avoid ruining the optimal solution.

A new implementation of BBO in [32] uses a simplified mutation strategy, which only mutates those habitats whose HSI values are low (e.g., solutions ranked in the last 20% in the population), and uses a unified mutation rate (e.g., 0.02) for each component of those habitats. This new strategy removes the requirements for calculating the species count probability and the mutation rate for each solution.

2.3.3 The Algorithmic Framework

Based on the migration and mutation operators, the framework of the basic BBO can be given by Algorithm 2.3.

Algorithm 2.3: The basic framework of BBO

---
1 Randomly initialize a population of N solutions (habitats);
2 **while** *stop criterion is not satisfied* **do**
3   Evaluate the fitness values of the solutions, based on which calculate their immigration, emigration, and mutation rates;
4   **for** $i = 1$ *to* N **do** perform migration on H_i according to Algorithm 2.1
5   **for** $i = 1$ *to* N **do** perform mutation on H_i according to Algorithm 2.2
6 **return** *the best solution found so far;*

In the implementation in [32], the migration operator does not modify the current solution; instead, it generates a new solution and then keeps the better one of the two solutions in the population. This method implies an elite strategy that the current best known solution will be either improved or retained for the next generation.

In practice, the number of species s in Eqs. (2.4) and (2.5) can be replaced by the value of fitness function f, so the migration rates can be calculated as

$$\lambda_i = I \frac{f_{max} - f(H_i) + \varepsilon}{f_{max} - f_{min} + \varepsilon} \tag{2.11}$$

$$\mu_i = E \frac{f(H_i) - f_{min} + \varepsilon}{f_{max} - f_{min} + \varepsilon} \tag{2.12}$$

Another way is to sort the solutions in decreasing order of their fitness values and then calculate the migration rates as

$$\lambda_i = I \frac{i}{n} \tag{2.13}$$

$$\mu_i = E \left(1 - \frac{i}{n}\right) \tag{2.14}$$

Simon [31] has given a recommended setting for BBO parameters as $N = 50$, $E = I = 1$, and $\pi_{max} = 0.01$. However, to achieve better performance, fine-tuning of the parameters is needed for each specific problem.

2.3.4 Comparison with Some Classical Heuristics

BBO is similar to many other heuristic algorithms in that it maintains a population of solutions and continuously evolves the solutions to search for the optimal solution (or some near-optimal solutions). However, BBO is distinct from other heuristics because of its theoretical model of biogeography and some other unique features.

In comparison with GA and other classical evolutionary algorithms, BBO does not "recombine" different individuals in the population, and its solutions are not treated as "parents" or "offsprings." BBO also does not explicitly use a selection operator, and each of its solutions is either maintained or improved from one iteration to the next. This differs from algorithms such as ACO that generates a new set of solutions at each iteration. This is also unlike algorithms such as evolutionary programming [12] which at each iteration combines parent' solutions and offsprings into a larger set, from which some solutions are picked out for the next generation.

Roughly speaking, BBO is more similar to PSO and DE where the initial population is maintained from one iteration to the next, and the solutions are continually evolved by self-adapting or learning from others. In PSO, solutions learn by moving toward the global best and the personal best; in DE, solutions learn by mixing difference with other solutions; while in BBO, solutions learn by immigrating components from others.

The work of Simon [31] has compared the performance of the basic BBO with a set of seven heuristic algorithms, including ACO [9], DE [35], PSO [6], evolution strategy (ES) [1], the basic GA [14] and two improved versions named PBIL [27] and SGA [19]. The test set consists of 14 benchmark function optimization problems from [4, 13, 44], and each algorithm is run for 100 times on each benchmark function. The experimental results show that BBO and SGA obtain the best average function values on seven functions, respectively, and their average performance is much better than the other fives. In terms of the minimum function value among the 100 runs, SGA, BBO, and ACO obtain the best results on seven, four, and three functions, respectively. Generally speaking, the overall performance of SGA is the best and BBO ranks the second. However, according to the no-free-lunch theorem [17, 42], we cannot conclude that some algorithms are absolutely superior to others, but the experiments demonstrate the good performance and potential of BBO.

2.4 Recent Advances of Biogeography-Based Optimization

In recent years, BBO has been widely studied and has achieved great success in the field of evolutionary computation. The studies can be divided into two major branches: The first is to modify the intrinsic mechanism of BBO, mainly for improving its search ability; the second is to extend BBO for different types of optimization problems. This section summarizes such existing studies of BBO. From the next chapter, we will begin to describe authors' work on BBO in detail.

2.4.1 Improved Biogeography-Based Optimization Algorithms

Blended BBO Migration is the core operator of BBO, and thus, early attempts to improve BBO focused on the model and/or the operation of migration. Ma and Simon [21] propose a hybrid BBO algorithm, called blended BBO (B-BBO) that borrows ideas from the crossover of GA [26]. It replaces the migration operator in the basic BBO with a blended migration operator, which mixes each component of an immigrating solution H_i with the corresponding component of the emigrating solution H_j as

$$H_i'(d) = \alpha H_i(d) + (1 - \alpha) H_j(d) \qquad (2.15)$$

where α is a real number between [0,1], which can be a predefined value (e.g., 0.5) or a random value. It is clear that the blended migration is a generalization of the original migration, and it becomes the original migration operator when $\alpha = 0$.

Fig. 2.3 Quadratic
migration model. Reprinted
from Ref. [20], ©2010, with
permission from Elsevier

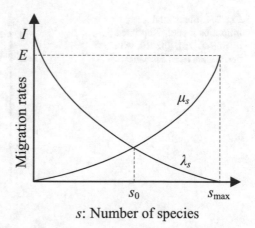

As the basic migration operator, the blended migration enables poor solutions
to improve themselves by migrating features from good solutions. However, unlike
the basic migration, the blended migration can partially avoid the degradation of
good solutions by keeping part of its original features, and it is capable of producing
new components different from either the immigrating and emigrating solutions to
improve diversity. Experimental results showed that the B-BBO performs better than
the basic BBO on a set of test functions [21].

Nonlinear Migration Models

There are different migration models in biogeography. The basic BBO uses the simplest linear migration model. However, the ecological system of nature is essentially
nonlinear, and small changes in a part of the system may cause large diffusion effects
on the whole system, so the actual species migration process is more complex than
the basic linear model. In [20], Ma introduced a variety of nonlinear migration models and selected some typical ones to study their influences on the performance of
BBO.

A typical nonlinear model is the quadratic migration model shown in Fig. 2.3,
where the migration rates are convex quadratic functions of the number of species:

$$\lambda_s = I\left(1 - \frac{s}{n}\right)^2 \tag{2.16}$$

$$\mu_s = E\left(\frac{s}{n}\right)^2 \tag{2.17}$$

The quadratic model is based on an experimentally tested theory of island biogeography [23], where the migration rate in a habitat follows a quadratic function
of the size of the habitat and the inner geographical proximity. When the habitat has
a small number of species, the immigration rate rapidly decreases from the maximum rate while the emigration rate slowly increases from zero. When the number

Fig. 2.4 Sinusoidal
migration model. Reprinted
from Ref. [20], ©2010, with
permission from Elsevier

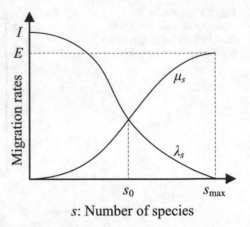

s: Number of species

of species in the habitat is nearly saturated, the decrease of the immigration rate becomes slower and the increase of the emigration rate becomes more rapid.

Another popular nonlinear model is the sinusoidal migration model shown in Fig. 2.4, where the migration rates are calculated as

$$\lambda_s = \frac{I}{2}\left(\cos\left(\frac{s\pi}{n}\right) + 1\right) \tag{2.18}$$

$$\mu_s = \frac{E}{2}\left(-\cos\left(\frac{s\pi}{n}\right) + 1\right) \tag{2.19}$$

The curves of the sinusoidal better match nature because it takes factors including predator/prey relationships, species mobility, evolution of species, population size, into consideration [41]. When the habitat has a small or large number of species, both the immigration rate and the emigration rate change slowly from their extremes. When the habitat has a medium number of species, the migration rates change rapidly from the equilibrium value, which means that the habitat needs to take a long time to reach the equilibrium again.

Ma [20] tests the effects of different migration models on the performance of BBO on 23 benchmark functions from [44], including high-dimensional unimodal functions (f_1–f_7), high-dimensional multimodal functions ($f8$–f_{13}), and low-dimensional multimodal functions (f_{14}–f_{23}). The BBO algorithms with different migration models used the same parameter settings as the basic BBO, and each algorithm is run for 50 times on each function. Table 2.1 presents the test results, which show that BBO with sinusoidal migration model outperforms the other two models on most of the functions, and the performance of BBO with quadratic migration model is also better than the basic BBO. The results indicate that the nonlinear model can better control the evolutionary process of the algorithm, and fine-tuning of the models is an effective way to improve the performance of BBO.

Table 2.1 Simulation results of the BBO algorithms using different migration models on 23 benchmark functions. Reprinted from Ref. [20], ©2010, with permission from Elsevier

f	Linear model			Quadratic model			Sinusoidal model		
	Best	Mean	Stdev	Best	Mean	Stdev	Best	Mean	Stdev
f_1	1.25E-03	5.23E-03	7.04E-03	7.14E-02	8.24E-02	7.03E-03	**1.02E-03**	**1.10E-03**	7.11E-04
f_2	7.23E-02	1.12E-01	3.54E-02	0.00E+00	0.00E+00	0.00E+00	**0.00E+00**	**0.00E+00**	0.00E+00
f_3	1.35E+02	1.43E+03	2.11E+02	6.41E+02	7.93E+03	6.56E+02	**2.04E+01**	**7.73E+02**	1.32E+02
f_4	5.53E-02	6.85E-02	8.93E-02	2.52E-15	4.31E-14	4.22E-14	**1.47E-15**	**7.56E-15**	4.32E-15
f_5	2.52E+00	3.32E+00	4.26E+00	**1.33E-02**	**5.14E-01**	7.45E-01	3.28E+00	8.76E+00	5.05E+00
f_6	1.24E+00	6.27E+00	3.74E+00	8.65E-03	1.19E-02	9.67E-02	**0.00E+00**	**0.00E+00**	0.00E+00
f_7	6.46E-03	8.32E-03	7.09E-03	9.01E-05	1.84E-04	4.35E-04	**3.77E-07**	**9.54E-07**	3.35E-07
f_8	3.42E-01	3.35E+00	3.41E-01	8.03E-01	2.72E+00	8.67E-01	**0.00E+00**	**0.00E+00**	0.00E+00
f_9	1.95E-01	1.03E+00	9.05E-01	**1.15E-02**	**1.56E-01**	1.23E-01	2.33E-01	4.50E-01	8.36E-01
f_{10}	2.45E-02	2.27E-01	2.10E-02	4.82E-01	5.38E-01	5.73E-01	**1.99E-02**	**2.12E-01**	2.05E-02
f_{11}	1.45E-01	6.78E-01	1.22E-01	2.76E-01	2.98E-01	3.89E-01	**1.02E-01**	**2.37E-01**	1.90E-02
f_{12}	7.27E-32	8.76E-32	9.66E-32	1.86E-32	2.87E-32	4.57E-31	**1.11E-32**	**1.71E-32**	6.89E-32
f_{13}	1.18E-03	2.10E-03	8.45E-03	**1.68E-33**	**5.54E-32**	1.99E-32	3.77E-32	9.03E-32	4.92E-32
f_{14}	1.41E-04	3.65E-04	3.21E-05	0.00E+00	0.00E+00	0.00E+00	**0.00E+00**	**0.00E+00**	0.00E+00
f_{15}	7.05E-04	8.90E-04	3.18E-05	**9.89E-09**	**5.89E-08**	6.92E-08	3.19E-04	5.29E-04	6.27E-05
f_{16}	4.12E-06	9.01E-06	9.22E-07	2.76E-08	3.85E-08	1.33E-08	**2.67E-09**	**1.51E-08**	7.32E-08
f_{17}	9.33E-03	5.05E-02	6.03E-02	8.11E-12	2.83E-10	7.55E-10	**2.17E-15**	**1.44E-14**	4.77E-14
f_{18}	6.51E-15	7.38E-15	9.04E-15	6.72E-12	6.04E-11	1.66E-11	**6.06E-15**	**7.05E-15**	2.56E-16
f_{19}	7.00E-25	8.77E-25	6.48E-25	3.62E-30	6.45E-30	3.35E-30	**0.00E+00**	**0.00E+00**	0.00E+00
f_{20}	9.26E-04	5.76E-03	5.09E-03	1.02E-08	4.65E-08	1.92E-09	**0.00E+00**	**0.00E+00**	0.00E+00
f_{21}	1.07E-03	2.38E-03	1.91E-03	1.33E-05	1.10E-04	2.43E-05	**5.29E-08**	**6.14E-07**	6.60E-08
f_{22}	4.77E-07	4.31E-06	7.87E-06	2.28E-09	3.27E-09	9.51E-10	**9.60E-12**	**2.89E-10**	7.31E-10
f_{23}	9.57E-08	8.65E-07	2.63E-07	4.35E-10	5.34E-09	6.37E-10	**3.55E-12**	**7.34E-11**	5.78E-12

Population Topologies For a population-based algorithm, the population topology also has a great impact on the performance of the algorithm. The authors have conducted an in-depth study on the population topologies of BBO and proposed a set of improved algorithms based on different local neighborhood structures [46]. Going one step further, we develop two new migration operators based on the local topologies and propose a novel ecogeography-based optimization (EBO) algorithm [46]. These new algorithms will be introduced in Chaps. 3 and 4, respectively.

Another research branch is to hybridize BBO with other heuristic algorithms, the studies of which will be described in Chap. 5.

2.4.2   Adaption of BBO for Constrained Optimization

A constrained optimization problem considers not only the objective function but also the constraints. The problem constraints divide the solutions into feasible solutions and infeasible solutions: A feasible solution satisfies all constraints, while an infeasible solution violates at least one constraint. There have been a variety of constraint-handling techniques [7, 24], among which the most popular one is the penalty function method. This method incorporates the constraints into the objective function by adding a penalty term to the objective function, so that the original constrained optimization problem is transformed into an unconstrained one.

According to the general form of constrained optimization problem shown in Eq. (1), a penalty function can be formulated as

$$\phi(\mathbf{x}) = \sum_{i=1}^{p} M_i |\min(0, g_i(\mathbf{x}))|^{\alpha} \tag{2.20}$$

where p is the number of constraints, g_i is the ith constraint, M_i are large positive constants, and α is a coefficient greater than or equal to 1. In many cases, we can simply use one constant M for all i and $\alpha = 1$. If the problem consists of less than inequality constraints $g_i(\mathbf{x}) \leq 0$ and equality constraints $g_i(\mathbf{x}) = 0$, the base terms can be set as $\max(0, g_i(\mathbf{x}))$ and $|g_i(\mathbf{x})|$, respectively.

Based on the penalty function, the original objective function f of the problem can be transformed to a penalized one as:

$$f'(\mathbf{x}) = f(\mathbf{x}) + \phi(\mathbf{x}) \tag{2.21}$$

Given that Eq. (1) represents a minimization problem, if a solution violates some constraints, its objective function will be much increased. For a maximization problem, $\phi(\mathbf{x})$ should be subtracted from $f(\mathbf{x})$.

The penalty function method is simple, but it is sensitive to the values of penalty factors M_i and α.

CBBO Boussaïd et al. [18] propose three variants of BBO, called CBBO, CBBO-DM, and CBBO-PM, for constrained optimization problems. These algorithms all adopt the penalty function method. They also use a stochastic ranking procedure [29] to dynamically balance the dominance of feasible and infeasible solutions. That is, each solution in the population is sorted based on both the fitness value and the amount of constraint violations. High ranking solutions have high emigration rates, and low ranking solutions have high immigration rates. The main difference among the three algorithms lies in their solution update strategies:

- CBBO uses the basic solution update strategy as the basic BBO.
- CBBO-DM replaces the original BBO mutation with the mutation of DE [28, 35] to generate new solution and then performs the migration on the mutated solutions.
- CBBO-PM replaces the original BBO mutation with a new mutation operator based on the pivot method [5], which generates a new mutated solution by combining the current best solution with another randomly selected solution.

Between each original solution H_i and its newly generated solution $H_{i'}$, the algorithm selects the better one into the next generation in terms of the following three cases:

- If both H_i and $H_{i'}$ are feasible, select the fitter one.
- If neither H_i nor $H_{i'}$ are feasible, select the one with the smaller penalty values.
- If one of H_i and $H_{i'}$ is feasible and the other is infeasible, select the feasible one.

The three variants are tested on 12 benchmark functions from [29] with constraints of four different types, including linear, nonlinear, equality, and inequality constraints. The experimental results show that the CBBO-DM performs better than the other two algorithms on most of the functions, which indicates that the DE mutation is better than the original BBO mutation and the PM method. By further comparing the effects of different DE mutation strategies, Boussaïd et al. suggest using the DE/rand/1/bin mutation strategies in CBBO-DM.

ε-BBO Bi and Wang [2] propose an ε-BBO algorithm that adopts the ε-constraint method [36] to deal with the constraints in the optimization problem. It also uses a new dynamic migration strategy to enhance the search ability of migration and employs a piecewise logistic chaotic to improve the population diversity.

As in the basic BBO, ε-BBO calculates the migration rates based on HSI. However, in ε-BBO the HSI ranking of each solution takes both the objective function and the constraint violation into consideration, so that the ranking of feasible solutions is improved. Algorithm 2.4 presents the procedure of ε for sorting all the N solutions in the population, where ϕ is the penalty function that calculates the constraint violation of each solution.

The dynamic migration strategy of ε-BBO incorporates a differential vector into the migration operation to produce deviant disturbance as follows:

$$H_i(d) = H_j(d) + (r_{\min} + \lambda_i \cdot r \cdot (r_{\max} - r_{\min})) \cdot (H_{r_1}(d) - H_{r_2}(d)) \qquad (2.22)$$

Algorithm 2.4: The sorting procedure of ε-BBO

---

1 **for** $i = 1$ *to* N **do**
2   **for** $j = 1$ *to* $N - 1$ **do**
3   **if** $(\phi(H_j) - \phi(H_{j+1}) \leq \varepsilon)$ *or* $(\phi(H_j) = \phi(H_{j+1}))$ **then**
4   **if** $(f(H_{j+1}) < f(H_j))$ **then** swap$(x(H_j), x(H_{j+1}))$
5   **else**
6   **if** $(\phi(H_{j+1}) < \phi(H_j))$ **then** swap$(x(H_j), x(H_{j+1}))$
7   **if** *no swap done* **then** break

---

where $H_i(d)$ and $H_j(d)$, respectively, denotes the dth dimension of the current solution H_i and the selected emigrating solution H_j, r_1 and r_2 are two random integer numbers between $[1, N]$, and the value of r increases with the iteration number t within the range of $[r_{min}, r_{max}]$ as

$$r^{(t)} = (1 - \beta) \cdot r^{(t)} + \beta \tag{2.23}$$

In this way, in the early stage of the search, the magnitude of disturbance is small, which is beneficial to reserve the original characteristics of promising solutions; in the later stage, the amplitude of disturbance increases in order to improve the diversity of the population.

ε-BBO also uses a new mutation strategy based on the piecewise logistic chaotic map [11], which exploits more effectively in the neighborhood around a local optimal solution to improve the solution accuracy. It first generates the chaotic sequence according to Eq. (2.24) and then performs the mutation as follows:

$$z_{t+1}(d) = \begin{cases} 4\mu z_t(d)(0.5 - z_t(d)), & 0 \leq z_t(d) < 0.5 \\ 4\mu(z_t(d) - 0.5)(1 - z_t(d)), & 0.5 \leq z_t(d) < 1 \end{cases} \tag{2.24}$$

$$H_i(d) = (2 - 2z_i(d)) \cdot H_i(d) + (2z_i(d) - 1) \times H_{best}(d) \tag{2.25}$$

where $z_t(d)$ is a random number between $(0,1)$, μ is a control parameter usually set to 4, and H_{best} denotes the current best solution.

Algorithm 2.5 presents the ε-BBO algorithm.

In [2], ε-BBO has been compared with B-BBO [21] on 13 benchmark constrained optimization functions $g1$–$g13$. The comparative result is that ε-BBO successfully obtains the global optima on 11 functions except g02 and g13, while B-BBO only obtains the global optima on 5 functions.

Algorithm 2.5: The ε-BBO algorithm

1 Initialize a population of N solutions;
2 **while** *the stop criterion is not satisfied* **do**
3   Sort the solutions based on Algorithm 2.4;
4   Calculate the immigration, emigration, and mutation rates of the solutions based on their ranking;
5   **for** *each solution H_i in the population* **do**
6   Perform dynamic migration according to Eqs. (2.22) and (2.23);
7   Perform piecewise logistic chaotic mutation according to Eqs. (2.24) and (2.25);
8   Compare the new solution with the original H_i by ε constrained method, and keep the winner in the population;
9   Update the best known solution;
10 **return** *the best solution found so far;*

2.4.3 Adaption of BBO for Multi-objective Optimization

The basic BBO is for solving single-objective optimization problems, where the HSI of a solution is calculated based on its objective function value. When there are multiple objectives, the calculation of HSI needs to take all the objective functions into consideration.

BBMO Ma et al. [22] proposed a biogeography-based multi-objective optimization (BBMO) algorithm, the main idea of which is to run multiple subprocedures of BBO in parallel, each of which for optimizing one objective function as in the basic BBO. During the parallel optimization process, the Pareto-order of each solution H_i is regarded as its fitness ranking $k(i)$, and thus migration rates are calculated as

$$\lambda_i = I\left(1 - \frac{k(i)}{n}\right) \tag{2.26}$$

$$\mu_i = E\frac{k(i)}{n} \tag{2.27}$$

MOBBO Bi et al. [3] propose a multi-objective optimization BBO algorithm, denoted as MOBBO, where the HSI of each solution is calculated according to its Pareto-dominance relation with other solutions in the population. Let $P = \{H_1, H_2, \ldots, H_N\}$ denote the population of N solutions, the fitness value of each H_i is calculated by Eq. (2.28), such that the fewer solutions dominated by those solutions that dominate H_i, the higher the fitness of H_i:

$$f(H_i) = 1 - \frac{1}{N} \sum_{H_j \in P \wedge H_j \prec H_i} |\{H_k \in P | H_j \prec H_k\}| \tag{2.28}$$

MOBBO then uses a self-adaptive method to calculate the migration rates of each H_i as follows (where α is the proportion of non-dominated solutions in P):

$$\lambda_i = \left(\frac{f_{\max} - f(H_i)}{f_{\max} - f_{\min}} \right)^{1/N^\alpha} \tag{2.29}$$

$$\mu_i = 1 - \lambda_i \tag{2.30}$$

At each generation, when the number of non-dominated solutions is larger than the population size N, MOBBO uses a competitive selection approach, first calculates the Euclidean distance between each pair of solutions, and then continually selects a pair of solutions whose distance is the smallest and removes the more crowded one until the number reaches N. This distance-based approach helps to maintain the diversity of the archive, which is important for multi-objective optimization since there are usually multiple Pareto-optimal solutions distributed in the search space. The basic framework of MOBBO is shown in Algorithm 2.6.

Algorithm 2.6: The MOBBO algorithm

---

1 Initialize a population P of N solutions and an empty archive A;
2 **while** *the stop criterion is not satisfied* **do**
3   Evaluate the fitness of the solutions according to Eq. (2.28);
4   Select all non-dominated solutions from **P** to **A**;
5   **if** $|A| \le N$ **then**
6   Remain the top N solutions in decreasing order of fitness in P;
7   **else**
8   **while** $|A| > N$ **do**
9   Calculate the matrix of pairwise distance of the solution in A;
10   Remove the more crowded one in the pair with the smallest distance from A;
11   Let $P = A$;
12   Calculate the migration rates of the solutions in **P** according to Eqs. (2.29) and (2.30);
13   Initialize an empty set P';
14   **forall the** H_i *in* P **do**
15   Perform dynamic migration according to Eqs. (2.22) and (2.23);
16   Perform piecewise logistic chaotic mutation according to Eqs. (2.24) and (2.25);
17   Add the resulting solution to P';
18   Let $P = P \cup P'$;
19 **return** *the non-dominated archive* A;

MOBBO is compared with some typical multi-objective evolutionary algorithms on the benchmark ZDT and DTLZ test problems, and the results show that MOBBO can obtain the smallest generational distance (GD) metric [37] and spacing (SP) metric [30] among the comparative algorithms, which demonstrates the fast convergence speed and good solution diversity achieved by MOBBO.

MO-BBO In [43], Xu proposes another multi-objective optimization algorithm based on BBO, denoted as MO-BBO, which uses the same fitness evaluation method as SPEA2 [47] and employs a new disturbance migration performed on the immigrating solution H_i as:

$$H_i'(d) = H_j(d) + \text{rand}() \cdot \left(H_{j_1}(d) - H_{j_2}(d)\right) \tag{2.31}$$

where H_j is the emigrating solution selected based on the emigration rate, while H_{j_1} and H_{j_2} are two other solutions randomly selected. If the resulting component is outside the predefined search range $[u_d, l_d]$ of the dth dimension, it is repaired by the following rule so that the component is set near the upper (lower) bound if it is originally closer to the upper (lower) bound:

$$H_i'(d) = \begin{cases} u_d - \text{rand}(0, 0.5) \cdot (u_d - H_i'(d)), & H_i'(d) > u_d \\ l_d + \text{rand}(0, 0.5) \cdot (l_d - H_i'(d)), & H_i'(d) < l_d \end{cases} \tag{2.32}$$

MO-BBO also uses an archive A of size M to maintain all current non-dominated solutions. At each generation, if the number of non-dominated solutions in $P \cup A$ is not larger than M, MO-BBO selects all of them into A of the next generation; otherwise, MO-BBO continually removes the most crowded non-dominated solution until the number is equal to M. Algorithm 2.7 presents the framework of MO-BBO.

Algorithm 2.7: The MO-BBO algorithm

---

1 Initialize a population P of N solutions and an empty archive A;
2 **while** *the stop criterion is not satisfied* **do**
3   Evaluate the fitness of the solutions according to Eq. (2.28);
4   Select non-dominated solutions from $P \cup A$ to A;
5   Calculate the migration rates of the solutions;
6   Initialize an empty set P';
7   **forall the** $H_i \subset \Lambda$ **do**
8   Perform the disturbance migration on H_i according to Eq. (2.32);
9   Perform the basic BBO mutation on the resulting H_i';
10   Add H_i' to P';
11   Combine P' and P to replace P;
12 **return** *the non-dominated archive A;*

2.4.4   Adaption of BBO for Combinatorial Optimization

Simon et al. [33] solve TSP by using a BBO algorithm, which produces new solutions (offsprings) by first selecting two parents p_1 and p_2, one with the probability

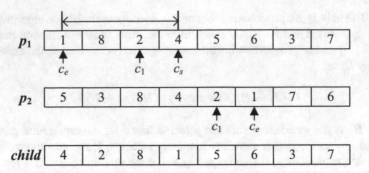

Fig. 2.5 Illustration of the inver-over operation for TSP

proportional to its immigration rate and the other with the probability proportional
to its emigration rate, and then employing the inver-over operation [38] performed
as follows:

(1) Randomly select a city c_1 from p_1, and then select the city following c_1 as the
    starting city c_s.
(2) Search for c_1 in p_2, and select the city following c_1 as the destination city c_e.
(3) Search for c_e in p_1, and reverse the sequence from c_s to c_e in p_1 to obtain the
    offspring.

Figure 2.5 illustrates the execution of the inver-over operator in a TSP instance
with eight cities. The experimental results show that the combination of BBO and the
inver-over operator is very effective in solving TSP [33]. Some other BBO algorithms
for TSP can be found in [25, 34].

For the 0-1 knapsack problem, Zhao et al. [45] propose a binary BBO algorithm,
which utilizes the migration operator of the basic BBO and designs two new mutation
operators. The first simply inverts the current component from 1 to 0 or vice versa, and
the second is a greedy mutation which is designed for infeasible solutions and which
always inverts a component with the minimum c_i/a_i until the solution becomes
feasible. The comparative experiments on four knapsack problem instances with
different sizes show that the binary BBO performs much better than GA in both
convergence speed and the success rate of finding out the optimum.

BBO has also shown its effectiveness on many other combinatorial optimization
problems, including the boxing problem [33], the clustering problem [15], the vertex
coloring problem [10]. The subsequent chapters also contain the applications of BBO
to a number of combinatorial optimization problems such as path planning and job
scheduling.

2.5 Summary

BBO is a relatively new heuristic algorithm, the key idea of which is to regard each solution as a habitat and evolve the solutions by enabling poor solutions to probably immigrate features from good solutions. This chapter introduces the basic principle, structure, and the recent advances of BBO. The remainder of the book will focus on the work done by the authors on the improvements and applications of BBO.

References

1. Beyer HG (2013) The theory of evolution strategies. Springer Science & Business Media, Berlin
2. Bi X, Wang J (2012) Constrained optimization based on epsilon constrained biogeography-based optimization. In: International conference on intelligent human-machine systems and cybernetics, pp 369–372. https://doi.org/10.1109/IHMSC.2012.184
3. Bi X, Wang J, Li B (2014) Multi-objective optimization based on hybrid biogeography-based optimization. Sys Eng Electron 179–186
4. Cai Z, Wang Y (2006) A multiobjective optimization-based evolutionary algorithm for constrained optimization. IEEE Trans Evol Comput 10:658–675
5. Clerc M (2003) Tribes-un exemple d'optimisation par essaim particulaire sans parametres de controle. In: Proceedings Sminaire Optimisation par Essaim Particulaire, pp 1–14
6. Clerc M (2006) Particle swarm optimization. ISTE. Wiley, New York
7. Coello Coello CA (2002) Theoretical and numerical constraint-handling techniques used with evolutionary algorithms: a survey of the state of the art. Comput Methods Appl Mech Eng 191:1245–1287. https://doi.org/10.1016/S0045-7825(01)00323-1
8. Darwin C (2004) On the origin of species, 1859. Routledge, London
9. Dorigo M, Di Caro G (1999) Ant colony optimization: a new meta-heuristic. In: Proceedings of the IEEE congress on evolutionary computation, vol 2, pp 1470–1477. https://doi.org/10.1109/CEC.1999.782657
10. Ergezer M, Simon D (2011) Oppositional biogeography-based optimization for combinatorial problems. In: Proceedings of the IEEE congress on evolutionary computation, pp 1496–1503
11. Fan J, Zhang X (2009) Piecewise logistic chaotic map and its performance analysis. Acta Electron Sin 720–725
12. Fogel DB (1994) An introduction to simulated evolutionary optimization. IEEE Trans Neural Netw 5:3–14. https://doi.org/10.1109/72.265956
13. Fogel DB (1997) Evolutionary algorithms in theory and practice. Complexity 2:26–27. https://doi.org/10.1002/(SICI)1099-0526(199703/04)2:4<26::AID-CPLX6>3.0.CO;2-7
14. Goldberg DE (1989) Genetic algorithms in search. Optimization and machine learning. Addison-Wesley, Reading
15. Hamdad L, Achab A, Boutouchent A (2013) Self organized biogeography algorithm for clustering. Nat Artif Models Comput Biol 396–405
16. Hanski I (1999) Metapopulation ecology. Oxford University Press, Oxford
17. Ho Y, Pepyne D (2002) Simple explanation of the no-free-lunch theorem and its implications. J Optim Theory Appl 115:549–570
18. Ilhem B, Amitava C, Patrick S, Mohamed AN (2012) Biogeography-based optimization for constrained optimization problems. Comput Oper Res 39:3293–3304. https://doi.org/10.1016/j.cor.2012.04.012
19. Khatib W, Fleming P (1998) The stud GA: a mini revolution? In: parallel problem solving from nature. Lecture notes in computer science, vol 1498, pp 683–691. https://doi.org/10.1007/BFb0056910

20. Ma H (2010) An analysis of the equilibrium of migration models for biogeography-based optimization. Inform Sci 180:3444–3464. https://doi.org/10.1016/j.ins.2010.05.035
21. Ma H, Simon D (2011) Analysis of migration models of biogeography-based optimization using markov theory. Eng Appl Artif Intell 24:1052–1060. https://doi.org/10.1016/j.engappai.2011.04.012
22. Ma H, Ruan X, Pan Z (2012) Handling multiple objectives with biogeography-based optimization. Int J Autom Comput 9:30–36
23. MacArthur R, Wilson E (1967) The theory of biogeography. Princeton University Press, Princeton
24. Michalewicz Z, Schoenauer M (1996) Evolutionary algorithms for constrained parameter optimization problems. Evol Comput 4:1–32
25. Mo H, Xu L (2010) Biogeography migration algorithm for traveling salesman problem. Adv Swarm Intell 405–414
26. Mühlenbein H, Schlierkamp-Voosen D (1993) Predictive models for the breeder genetic algorithm I. continuous parameter optimization. Evol Comput 1:25–49. https://doi.org/10.1162/evco.1993.1.1.25
27. Parmeec IC (2001) Evolutionary and adaptive computing in engineering design. Springer Science & Business Media, Berlin
28. Price K, Storn RM, Lampinen JA (2005) Differential evolution: a practical approach to global optimization. Natural computing series
29. Runarsson T, Yao X (2000) Stochastic ranking for constrained evolutionary optimization. IEEE Trans Evol Comput 4:284–294. https://doi.org/10.1109/4235.873238
30. Schott JR (1995) Fault tolerant design using single and multicriteria genetic algorithm optimization. Cell Immunol 37
31. Simon D (2008) Biogeography-based optimization. IEEE Trans Evol Comput 12:702–713. https://doi.org/10.1109/TEVC.2008.919004
32. Simon D (2009) Biogeography-based optimization. http://academic.csuohio.edu/simond/bbo/
33. Simon D, Rarick R, Ergezer M, Du D (2011) Analytical and numerical comparisons of biogeography-based optimization and genetic algorithms. Inform Sci 181(7):1224–1248
34. Song Y, Liu M, Wang Z (2010) Biogeography-based optimization for the traveling salesman problems. In: Third international joint conference on computational science and optimization, vol 1, pp 295–299. https://doi.org/10.1109/CSO.2010.79
35. Storn R, Price K (1997) Differential evolution - a simple and efficient heuristic for global optimization over continuous spaces. J Global Optim 11:341–359. https://doi.org/10.1023/A:1008202821328
36. Takahama T, Sakai S (2010) Efficient constrained optimization by the constrained adaptive differential evolution. In: IEEE congress on evolutionary computation, pp 1–8. https://doi.org/10.1109/CEC.2010.5586545
37. Tang L, Wang X (2013) A hybrid multiobjective evolutionary algorithm for multiobjective optimization problems. IEEE Trans Evol Comput 17:20–45
38. Tao G, Michalewicz Z (1998) Inver-over operator for the TSP. Parallel Probl Solving Nat 803–812
39. Wallace A (1876) The geographical distribution of animals: with a study of the relations of living and extinct faunas as elucidating the past changes of the earth's surface. Macmillan and Company
40. Wesche TA, Goertler, CM, Hubert WA (1987) Modified habitat suitability index model for brown trout in Southeastern Wyoming. N Am J Fish Manag 7:232–237
41. Whittaker RJ, Fernández-Palacios JM (1998) Island biogeography. Oxford University Press, Oxford
42. Wolpert D, Macready W (1997) No free lunch theorems for optimization. IEEE Trans Evol Comput 1:67–82. https://doi.org/10.1109/4235.585893
43. Xu Z (2013) Study on multi-objective optimization theory and application based on biogeography algorithm. Ph.D. thesis

44. Yao X, Liu Y, Lin G (1999) Evolutionary programming made faster. IEEE Trans Evol Comput 3:82–102. https://doi.org/10.1109/4235.771163
45. Zhao B, Deng C, Yang Y (2012) Novel binary biogeography-based optimization algorithm for the knapsack problems. Advances in swarm intelligence. Lecture notes in computer science, vol 7331. Springer, Berlin, pp 217–224
46. Zheng YJ, Ling HF, Xue JY (2014) Ecogeography-based optimization: enhancing biogeography-based optimization with ecogeographic barriers and differentiations. Comput Oper Res 50:115–127. https://doi.org/10.1016/j.cor.2014.04.013
47. Zitzler E, Laumanns M, Thiele L (2001) SPEA2: improving the strength pareto evolutionary algorithm for multiobjective optimization. In: Evolutionary methods for design, optimisation and control with applications to industrial problems. https://doi.org/10.3929/ethz-a-004284029

Chapter 3
Localized Biogeography-Based Optimization: Enhanced By Local Topologies

3.1 Introduction

For heuristic algorithms that evolve a population of solutions, the population topology defines how the solutions interact with each other in the population, which often has a great influence on the performance of the algorithms. This chapter first introduces some typical types of population topologies and then describes the methods for improving BBO with local topologies.

3.2 Population Topology

The population topology of an algorithm defines the neighborhood $\mathbf{N}(X)$ for each solution X in the population, i.e., the set of solutions X can interact with. In general, the larger the average neighborhood size of all solutions, the faster the information transfers among the solutions, and the faster the algorithm converges. However, a large neighborhood size can also incur more computational cost and make the algorithm be more likely to be trapped in local optimum. So, it is important to choose a suitable topology in algorithm design.

3.2.1 Global Topology

In the global topology, all solutions in the population are interconnected, and information transferring can occur between any two solutions, as illustrated in Fig. 3.1. In generally, a population-based algorithm uses the global topology by default. For example, in classical GA, any two chromosomes have chances to participate a crossover operation; in classical ACO, the pheromone left by an ant may affect any

© Springer Nature Singapore Pte Ltd. and Science Press, Beijing 2019
Y. Zheng et al., *Biogeography-Based Optimization: Algorithms and Applications*,
https://doi.org/10.1007/978-981-13-2586-1_3

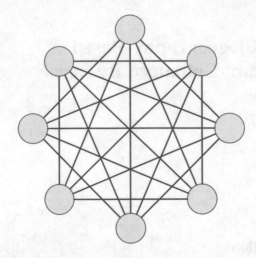

Fig. 3.1 Global topology, where the vertices denote solutions and the edges denote connections between the solutions

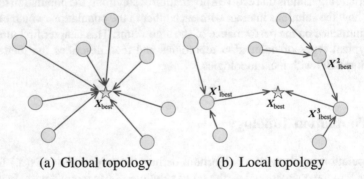

| (a) Global topology | (b) Local topology |

Fig. 3.2 Illustration of interactions among solutions in global and local topologies, where X_{best} denotes the global best and X_{best} denotes a local best

other ant. Suppose the population size is N, in a global topology each solution can interact with the other $(N-1)$ ones, and thus the neighborhood size is $(N-1)$.

Most of heuristic algorithms have some mechanisms to enable their solutions to learn from the current best solution. For example, in PSO the particles move by learning from **gbest**; in GA, the global best has the highest probability of being selected and producing offspring. Under the global topology, if the global best is much better than the others while itself is trapped by a local optima, then all the solutions are very likely to move toward the optima, and the algorithm is easy to converge prematurely, as shown in Fig. 3.2a.

Fig. 3.3 Isolated topology, where each subset of then population is marked with a dotted circle

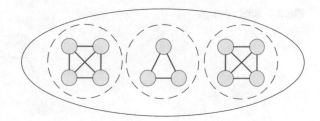

3.2.2 Local Topologies

In a local topology, every solution is only connected with a part of other solutions, and thus, the information transferring in the population and the convergence speed of the algorithm will be slower than in a global topology.

Note that a "slow" convergence speed does not always means "bad", as it decreases the probability of being trapped in local optima. In a local topology, every solution learns from the best one in its neighborhood (called the local best), as shown in Fig. 3.2b. In this way, although if one or more local bests approach some local optima, many others solutions are not affected and can continue to search in other regions of the search space, and thus, the algorithm is not easy to be trapped in local optima.

Here, we introduce some typical local topologies.

Isolated topology The isolated topology simply isolates the population into a number of subsets, the solutions in the same subset can interact with each other, and there is no interaction between any two solutions from two different subsets, as shown in Fig. 3.3. Suppose the number of the subsets is K, the neighborhood size of the isolated topology is $(N/K) - 1$.

The isolated topology is easy to implement, as it only needs to set a number K of the subsets to partition the solutions. It is easy to see that the neighborhood size of the isolated topology is $(N/K) - 1$. As every subset can explore a different region of the search space, the algorithm is not easy to be trapped in a local optimum. However, as there is no communication between different subsets, those subsets searching in inferior regions will contribute little to the algorithm, and thus, the efficiency is often low.

Ring topology The ring topology is another simplest topology, where the solutions are assumed to be connected one by one in a ringlike mode, and each solution can only interact with two solutions before and after it, as shown in Fig. 3.4. It is obvious that the neighborhood size of the ring topology is 2, and its speed of information transferring is often very slow.

In algorithm implementation, suppose all the N solutions are maintained in an array or a list, then the two neighbors of each solution X_i can be obtained as $X_{(i-1)\%n}$ and $X_{(i+1)\%n}$, where % denotes the modulo operation.

Square topology In the square topology, the solutions are arranged in a grid-like way, and each solution has four immediate neighbors, as shown in Fig. 3.5 (Note that

Fig. 3.4 Ring topology

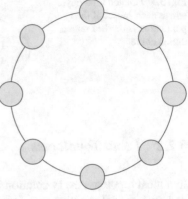

Fig. 3.5 Square topology

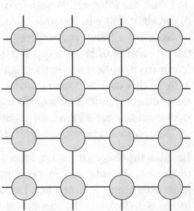

the solutions in the top line are connected with the corresponding solutions in the bottom line, so are the solutions in left line and the corresponding ones in right line, which are not represented directly by the figure). It is obvious that the neighborhood size of a square topology is 4, and its speed of information transferring is faster than that in the ring topology.

In algorithm implementation, suppose that the population size $N = K^2$, then the indices of the four neighbors of each solution X_i can be obtained as $(i - K)\%N$, $(i + K)\%N$, $(i + K_1)$, and $(i - K_2)$, respectively, where K_1 is $(K - 1)$ if $i\%K = 0$ and is -1 otherwise, and K_2 is $(K - 1)$ if $(i + 1)\%K = 0$ and is 1 otherwise.

Star topology As shown in Fig. 3.6a, in the star topology, a central solution is connected to all other ones in the population, but each other solution is only connected to the central solution. It is obvious that the neighborhood size is $2(N - 1)/N$ which is even less than 2 of the ring topology.

The search process of an algorithm using the star topology is affected greatly by the central solution and is easy to be trapped in a local optimum if the central solution approaches the local optimum. There are some variants of the star topology

Fig. 3.6 Star topology and its variant

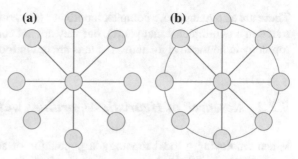

Fig. 3.7 Diamond topology

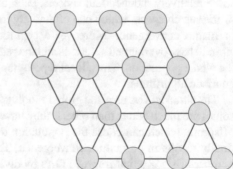

by allowing connections between some non central solutions. For example, as shown in Fig. 3.6b, the non-central solutions are connected one after another, which can be regarded as a combination of the star topology and the ring topology.

Diamond topology As shown in Fig. 3.7, in a diamond topology, each solution has six immediate neighbors, and the neighborhood size is 6. It is often used when we need a relatively fast convergence speed, but not so fast as the global topology.

Random topology In the local topologies mentioned above, the neighbors of each solution are fixed. On the contrary, the random topology randomly chooses for each solution a set of neighbors. There are two popular approaches for setting a random topology:

- Given a neighborhood size K_N, for each solution X in the population, randomly select other K_N neighbors for X. By this way, all solutions have the same neighborhood size.
- Given an average neighborhood size K_N, for each solution X in the population, each other solution has a probability of $K_N/(N-1)$ of being the neighbor of K. The approach is more "random" than the former.

The setting of the random topology can be arbitrary. In many algorithms using the random topology, during the search process (e.g., after some iterations of the algorithm) the neighborhood structure can be reset; that is, the neighbors of each solution are randomly selected again.

There are also many more complex topologies (some of which even cannot be illustrated in a two-dimensional plane), but they are not commonly used for population topology in heuristic algorithms and thus are not introduced here.

3.2.3 Research of Heuristic Algorithms with Local Topologies

When employing a local topology, a population of solutions are partitioned into some relatively independent subsets, each of which exploring in a different region of the search space, which can effectively improve the solution diversity and avoid premature convergence. Moreover, except for the isolated topology, different subsets can still have overlapped solutions and thus enable information to flow slowly through the whole population. Thus, local topologies provide an effective way for improving heuristic algorithms.

The classical GA uses the global topology by default. So, if one or more chromosomes are much better than others, they have greater chances than others to produce offspring, which can cause the population to be more and more homogeneous and finally results in premature convergence. To tackle with this problem, there are a number of studies that improve GAs by dividing the population different subpopulations according to the similarity of the individuals in the population and evolving these subpopulations independently [5]. This is inspired by the nature where different species evolve to fill different subspaces (called niches) in the environment that can support different types of life. A "niching" GA typically uses a sharing function that degrades an individual's fitness based on the presence of other neighboring individuals and thus provides enough diversification for solutions in the population and suppresses premature convergence [22]. However, static niching methods for subpopulation division can significantly restrict information sharing of the population and the convergence speed of the algorithm. One improved strategy is to use dynamical niching, i.e., redivide subpopulations during the search process to break the isolation [13]. Another advantage of niching methods is that different subpopulations can perform independent search for multiple global optima or very good local optima that are needed by the users. Thereby, niching methods have been achieved increasing interests for multimodal optimization and multi-objective optimization [6, 12, 13, 21].

The problem of premature convergence is also serious in the basic PSO algorithm where all particles in the swarm learn from the same global best. Shortly after the proposal of the basic PSO, Kennedy [9] introduces the local topology into PSO, where each particle learns from the local best in its neighborhood instead of the global best to alleviate premature convergence. In [10], Kennedy and Mendes conduct a systematic research on PSO with different local topologies, the result of which indicates that the effect of the local topologies on algorithm performance varies from problems to problem. Li [11] proposes a PSO algorithm combining the ring topology with niching to simplify parameters setting and enhance information sharing. Later studies show that PSO using dynamic local topologies typically exhibit better performance on many problems [1, 7, 15]. In particular, the standard PSO versions published on www.particleswarm.info (the latest one is SPSO2011) [3] all use the dynamical random topology.

For DE algorithms, Chakraborty et al. [2] and Omran et al. [16], respectively, find that replacing the global best in DE/best/1/bin and DE/best/2/bin schemes with the local best based on local topologies can effectively suppress premature. Das et al. [4] further propose a DE algorithm that combines the local and global topologies and uses a weight parameter to control which topology should be employed. The strategy of combined topologies has exhibited a promising performance on a variety of multimodal optimization problems [18].

Local topologies have also been successfully applied to many novel heuristic algorithms. Pan et al. [17] introduce the local topology into HS algorithm, which make each new harmony has a high probability of learning from the best neighbor based on a dynamic topology. Mo and Zeng [14] test the use of some simple local topologies in artificial physics optimization algorithm. Hu et al. [8] employ the ring topology in a memetic algorithm for constrained optimization. Most studies show that in most cases, a heuristic algorithm using an appropriate local topology can be superior to the basic algorithm with the global topology.

3.3 Localized Biogeography-Based Optimization Algorithms

The basic BBO uses the global topology, where any two habitats may interact with each other: If a habitat is to be immigrated, any other habitat have a chance to be the emigrating habitat. The algorithm is easy to be trapped in local optima if one or more habitats which are at the local optima have high emigration rates.

To tackle with this problem, we first introduce the ring topology into BBO [24], and then conduct a more comprehensive research on BBO with local topologies (which we call Local-BBO algorithms) [25]. The experiment results demonstrate that the local topologies can effectively improve the performance of BBO.

3.3.1 Local-BBO with the Ring Topology

We propose a BBO algorithm with the ring topology, denoted as RingBBO. Suppose there are N solutions in a population implemented as an array $\mathbf{H}$, in which the ith solution is denoted by H_i. As described in Sect. 3.2.2, using the ring topology, each solution H_i has only directly connected with its left neighbor $H_{(i-1)\%N}$ and its right neighbor $H_{(i+1)\%N}$, the indices of which are denoted by $i1$ and $i2$, respectively. When H_i is to be immigrated, the probability of selecting the left neighbor as the emigrating solution is $\mu_{i1}/(\mu_{i1} + \mu_{i2})$, and the probability of selecting the right neighbor is $\mu_{i2}/(\mu_{i1} + \mu_{i2})$.

Algorithm 3.1 presents the migration procedure under the ring topology, where D denotes the dimension of problems.

Algorithm 3.1: The migration procedure under the ring topology

---
1 **for** $d = 1$ *to* D **do**
2   **if** rand() $< \lambda_i$ **then**
3   Let $i1 = (i - 1)\%N$, $i2 = (i + 1)\%N$;
4   **if** rand() $< \mu_{i1}/(\mu_{i1} + \mu_{i2})$ **then** set $H_i(d) = H_{i1}(d)$
5   **else** set $H_i(d) = H_{i2}(d)$

By replacing "Algorithm 2.1" with "Algorithm 3.1" in Line 4 of Algorithm 2.3 in Chap. 2, we get the algorithmic framework of RingBBO.

Note that in the basic BBO, each migration operation needs to select the emigrating solution with a probability proportional to the emigration rate from the whole population, and thus, the time complexity of the operation is $O(N)$. However, using the ring topology, it just needs to select either the left neighbor or the right neighbor as the emigrating solution, and thus, the time complexity is only $O(2)$. Thereby, the complexity of each iteration of the basic BBO is $O(N^2 D)$, while the complexity of each iteration of RingBBO reduces to $O(2ND)$.

3.3.2 Local-BBO with the Square Topology

Using the square topology in BBO (denoted as SquareBBO), at each migration operation, the emigrating solution is selected from four neighbors of the immigrating solution. Let μ_1, μ_2, μ_3, and μ_4 be the emigration rates of the four neighbors, respectively, and then their probabilities of being the emigrating solution are $\mu_1/\left(\sum_{i=1}^{4} \mu_i\right)$, $\mu_2/\left(\sum_{i=1}^{4} \mu_i\right)$, $\mu_3/\left(\sum_{i=1}^{4} \mu_i\right)$ and $\mu_4/\left(\sum_{i=1}^{4} \mu_i\right)$, respectively. Algorithm 3.2 presents the migration procedure under the square topology.

Similarly, by replacing "Algorithm 2.1" with "Algorithm 2.3" in Line 4 of Algorithm 2.3, we get the algorithmic framework of SquareBBO. The time complexity of a migration operation is $O(4)$; thus, the time complexity of each iteration of SquareBBO is $O(4ND)$.

3.3.3 Local-BBO with the Random Topology

Using the random topology, we can set the neighborhood size to an arbitrary K_N between 0 and $(N-1)$, which can be adjusted according to the properties of the

---

Algorithm 3.2: The migration operation under the square topology

---

1 **for** $d = 1$ *to* D **do**
2   **if** rand() $< \lambda_i$ **then**
3   let $K = \sqrt{N}$;
4   let $i1 = (i - K)\%N, i2 = (i + K)\%N$ let $i3 = i + K - 1$ if $i\%K = 0$ else $i - 1$;
5   let $i4 = i - K + 1$ if $(i + 1)\%K = 0$ else $i + 1$;
6   let $r = $ rand() $\cdot (\mu_{i1} + \mu_{i2} + \mu_{i3} + \mu_{i4})$;
7   **if** $r < \mu_{i1}$ **then** set $H_i(d) = H_{i1}(d)$;
8   **else if** $r < (\mu_{i1} + \mu_{i2})$ **then** set $H_i(d) = H_{i2}(d)$; **else if** $r < (\mu_{i1} + \mu_{i2} + \mu_{i3})$ **then** set $H_i(d) = H_{i3}(d)$;
9   **else** set $H_i(d) = H_{i4}(d)$

---

problem. Thus, the random topology is much more flexible than other topologies with neighborhood sizes (such as 2 in the ring topology and 4 in the square topology).

As mentioned above, there are two popular approaches for setting a random topology. The first is to randomly select other K_N neighbors for each solution, but it has the following two restrictions:

- K must be an integer, which limits the adjustability of the parameter.
- All solutions have the same number of neighbors, which is not reasonable for some applications. For example, for a model abstracted from a social network, different individuals usually have different numbers of neighbors.

The second is to let each pair of solutions has a probability of $K_N/(N-1)$ of being connected, which not only increases the adjustability of the parameter, but also enriches the diversity of the neighborhood structure. More precisely, this approach gives a *Bernoulli* distribution where any size between 0 and $(N-1)$ is possible, such that all possible topologies may appear. Therefore, we employ the second approach in our BBO algorithm with the random topology, denoted as RandBBO.

In algorithm implementation, the neighborhood structure can be maintained by an $N \times N$ matrix **Link**, where **Link**$(i, j) = 1$ indicates that two solutions H_i and H_j are connected and **Link**$(i, j) = 0$ otherwise $(1 \leq i, j \leq N)$. Algorithm 3.3 shows the procedure for setting the random topology with an expected neighborhood size of K_N:

Algorithm 3.3: Set the adjacent matrix for the random neighborhood structure with an expected size of K_N

---

1 Initialize an $N \times N$ matrix **Link**;
2 Let $p = K_N/(N - 1)$;
3 **for** $i = 1$ *to* N **do**
4   **for** $j = 1$ *to* N **do**
5   **if** rand() $< p$ **then** set **Link**$(i, j) = $ **Link**$(j, i) = 1$;
6   **else** set **Link**$(i, j) = $ **Link**$(j, i) = 0$;

Employing roulette wheel selection, the migration operation under the random topology can be described in Algorithm 3.4.

Algorithm 3.4: The migration operation under the random topology

---

1 **for** $d = 1$ *to* D **do**
2   **if** rand() $< \lambda_i$ **then**
3   Let $J = \{H_j \in \mathbf{P} | j \neq i \wedge \mathbf{Link}(i, j) = 1\}$;
4   Let $s = \sum_{H_j \in J} \mu_j$;
5   let $k = 1, c = \mu_{J(1)}, r = \text{rand}() \cdot s$;
6   **while** $c < r \wedge k \leq |J|$ **do**
7   Set $k = k + 1, c = c + \mu_{J(k)}$;
8   Set $j = J(k), H_i(d) = H_j(d)$;

The time complexity of the migration under the random topology is $O(K_N)$; thus, the complexity of each iteration of the algorithm is $O(K_N N D)$.

Dynamically resetting the neighborhood structure can enrich the diversity and improve the search efficiency of the algorithm. In RandBBO, we use a simple strategy that resets the neighborhood structure if the algorithm does not find any new best solution after a given number $\hat{I}$ of iterations. Algorithm 3.5 presents the framework of RandBBO.

Algorithm 3.5: The RandBBO algorithm

---

1 Randomly initialize a population of solutions and let $I = 0$;
2 Evaluate the fitness of the solutions and select the best one as H_{best};
3 Initialize the neighborhood structure according to Algorithm 3.3;
4 **while** *the stop criterion is not satisfied* **do**
5   Calculate the migration and mutation rates of the solutions;
6   **for** $i = 1$ *to* N **do**
7   Perform the migration on H_i according to Algorithm 3.4;
8   Perform the mutation on H_i according to Algorithm 2.2;
9   Evaluate the fitness of the resulting H_i';
10   **if** $f(H_i) > f(H_i')$ **then**
11   Use H_i' to replace H_i in the population;
12   **if** $f(H_i') > f(H_{best})$ **then**
13   set $H_{\text{best}} = H_i'$;
14   set $I = 0$;

15   **if** *No new best solution is found* **then**
16   set $I = I + 1$;
17   **if** $I = \hat{I}$ **then**
18   Reset the neighborhood structure according to Algorithm 3.3;
19   set $I = 0$;

20 **return** H_{best};

3.4 Computational Experiments

We compare the three versions of Local-BBO with the basic BBO on a set of 23 benchmark functions from [23], denoted as f_1-f_{23}, which span a diverse set of features of numerical optimization problems and are widely used in the areas of continuous optimization. f_1-f_{13} are high-dimensional and scalable problems, and $f_{14}-f_{23}$ are low-dimensional problems with a few local minima. In the first thirteen high-dimensional functions, f_1-f_5 are unimodal functions, f_6 is a step function, which only has a minimum and is not continuous, f_7 is a quartic function containing noise, f_8-f_{13} are multimodal functions, whose local minimum can increase exponentially with the dimension of problems, and the kind of these problems is hard to solve for most optimization algorithms. Table 3.1 presents essential information of the benchmark functions, where x^* is the optimal solution and $f(x^*)$ is the optimal objective function value. As we can see, for all high dimensional except f_8, the optimal function values are zero, and f_8 and all the low-dimensional functions have nonzero optimal values. For convenience, in the experiment we add some constants to those functions such that their optimal values become zero.

The basic BBO is implemented as in [19], where the parameters are set as $N = 50$, $I = E = 1$, and $\pi_{max} = 0.01$. For the three Local-BBO algorithms, we set the same population size and maximum migration rates, but set $\pi_{max} = 0.02$ to increase the chance for mutation since the local topologies decrease the impact of migration. For RandBBO, we set $K_N = 3$ and $\hat{I} = 1$. The stop criterion is that the number of function evaluations (NFE) reaches a predefined maximum NFE (MNFE). For every problem, we set a same MNFE (shown in column 2 of Table 3.2) for all the algorithms for a fair comparison. Every algorithm is run on each problem for 60 times with different random seeds, and the resulting function values are averaged over 60 runs. The experimental environment is a computer of Intel Core i5-2520M processor and 4 GB memory.

Table 3.2 presents the experimental results. In columns 3–6, the upper part of each cell denotes the mean best value obtained by the algorithm and lower part denotes its standard deviation. As we can see, the mean bests of three Local-BBO algorithms are always better than the basic BBO on all the functions except f_7. In particular, on all the high-dimensional functions except f_7, the mean bests of Local-BBO are only about 1% ∼ 20% of that of the basic BBO, which demonstrates the significant improvement brought by the local topologies.

In addition, we use paired t-tests to identify statistical differences of the results. In columns 3–6 of Table 3.2, a superscript "+" means that the corresponding result of Local-BBO has a statistically significant improvement over the basic BBO, and "−" means vice versa (at 95% confidence level). The last row of the table summarizes the number of functions on which the Local-BBO significantly outperforms the basic BBO versus vice versa. The results show that, RingBBO and SquareBBO are significantly better than the basic BBO on 20 functions, and RandBBO does so on 21 functions. There are no significant differences among the four algorithms on function f_{17}, which is low dimensional and is one of simplest functions among the test

Table 3.1 A summary of the 23 benchmark functions

No.	Function	Search range	D	x^*	$f(x^*)$
f_1	Sphere	$[-100, 100]^D$	30	0^D	0
f_2	Schwefel 2.22	$[-10, 10]^D$	30	0^D	0
f_3	Schwefel 1.2	$[-100, 100]^D$	30	0^D	0
f_4	Schwefel 2.21	$[-100, 100]^D$	30	0^D	0
f_5	Rosenbrock	$[-2.048, 2.048]^D$	30	1^D	0
f_6	Step	$[-100, 100]^D$	30	0^D	0
f_7	Quartic	$[-1.28, 1.28]^D$	30	0^D	0
f_8	Schwefel	$[-500, 500]^D$	30	420.9687^D	-12569.5
f_9	Rastrigin	$[-5.12, 5.12]^D$	30	0^D	0
f_{10}	Ackley	$[-32.768, 32.768]^D$	30	0^D	0
f_{11}	Griewank	$[-600, 600]^D$	30	0^D	0
f_{12}	Penalized1	$[-50, 50]^D$	30	1^D	0
f_{13}	Penalized2	$[-50, 50]^D$	30	1^D	0
f_{14}	Shekel's Foxholes	$[-65.536, 65.536]^D$	2	$(-32, 32)$	1
f_{15}	Kowalik	$[-5, 5]^D$	4	$(0,1928,0.1908, 0.1231,0.1358)$	0.0003075
f_{16}	Six-Hump Camel-Back	$[-5, 5]^D$	2	$(\pm 0.08983, \mp 0.7126)$	-1.0316285
f_{17}	Branin	$[-5, 10] \times [0, 15]$	2	$(\pm 3.142, 7.252 \mp 5)$	0.398
f_{18}	Goldstein-Price	$[-2, 2]^D$	2	$(0, -1)$	3
f_{19}	Hartman's Family 1	$[0, 1]^D$	3	$(0.114, 0.556, 0.852)$	-3.86
f_{20}	Hartman's Family 2	$[0, 1]^D$	6	$(0.201, 0.150, 0.477), (0.275, 0.311, 0.657)$	-3.32
f_{21}	Shekel's Family 1	$[0, 10]^D$	4	5 optima	-10.1532
f_{22}	Shekel's Family 2	$[0, 10]^D$	4	7 optima	-10.4029
f_{23}	Shekel's Family 3	$[0, 10]^D$	4	10 optima	-10.5364

problems. The basic BBO is significantly better only on f_7, which is a noisy function with one minima. For such a simple function, the basic BBO with global opology is more capable of tolerating noises. However, since the RingBBO and SquareBBO have low time complexity, when given the same computational time, we observe that RingBBO and SquareBBO still achieve better results than the basic BBO. In most cases, the local topologies have much more advantages.

For each problem, we also set a required accuracy, which is 10^{-2} for f_2 and 10^{-8} for the other 22 functions [20]: If the error of best objective function value found by

Table 3.2 Experimental results of the basic BBO and Local-BBO on the 13 test functions

f	MNFE	BBO	RingBBO	SquareBBO	RandBBO
f_1	150,000	2.43E+00 (8.64E–01)	2.90E–01 $^+$ (1.01E–01)	3.09E–01 $^+$ (1.19E–01)	3.26E–01 $^+$ (1.38E–01)
f_2	150,000	5.87E–01 (8.60E–02)	1.96E–01 $^+$ (3.72E–02)	1.95E–01 $^+$ (4.18E–02)	2.09E–01 $^+$ (3.73E–02)
f_3	500,000	4.23E+00 (1.73E+00)	4.88E–01 $^+$ (2.09E–01)	4.09E–01 $^+$ (1.58E–01)	3.98E–01 $^+$ (1.58E–01)
f_4	500,000	1.62E+00 (2.80E–01)	7.61E–02 $^+$ (1.00E–01)	6.55E–01 $^+$ (1.44E–01)	7.00E–01 $^+$ (1.25E–01)
f_5	500,000	1.15E+02 (3.66E+01)	8.60E+01 $^+$ (2.84E+01)	6.63E+01 $^+$ (3.37E+01)	7.83E+01 $^+$ (3.25E+01)
f_6	150,000	2.20E+00 (1.61E+00)	0.00E+00 $^+$ (0.00E+00)	3.33E–02 $^+$ (1.83E–01)	3.33E–02 $^+$ (1.83E–01)
f_7	300,000	3.31E–03 (3.01E–03)	6.16E–03 $^-$ (3.65E–03)	5.93E–03 $^-$ (3.13E–03)	5.90E–03 $^-$ (2.44E–03)
f_8	300,000	1.90E+00 (8.02E–01)	2.44E–01 $^+$ (1.17E–01)	2.35E–01 $^+$ (9.27E–02)	2.04E–01 $^+$ (5.92E–02)
f_9	300,000	3.51E–01 (1.40E–01)	3.98E–02 $^+$ (1.73E–02)	3.94E–02 $^+$ (1.18E–02)	3.69E–02 $^+$ (1.06E–02)
f_{10}	150,000	7.79E–01 (2.03E–01)	1.76E–01 $^+$ (4.02E–02)	1.77E–01 $^+$ (4.85E–02)	1.63E–01 $^+$ (4.21E–02)
f_{11}	200,000	8.92E$^+$01 (1.01E–01)	2.36E–01 $^+$ (7.00E–02)	2.35E–01 $^+$ (8.34E–02)	2.82E–01 $^+$ (8.34E–02)
f_{12}	150,000	7.35E–02 (2.87E–02)	7.32E–03 $^+$ (3.68E 03)	6.29E 03 $^+$ (4.23E–03)	7.71E–03 $^+$ (7.19E–03)
f_{13}	150,000	3.48E–01 (1.23E 01)	3.81E–02 $^+$ (2.81E–02)	4.19E–02 $^+$ (2.87E–02)	3.90E–02 $^+$ (2.71E–02)
f_{14}	50,000	3.80E–03 (1.01E–02)	3.88E–06 $^+$ (1.41E–05)	3.44E–06 $^+$ (1.87 – 05)	1.01E–06 $^+$ (4.04E–06)
f_{15}	50,000	6.59E–03 (7.70E–03)	2.42E–03 $^+$ (2.12E–03)	3.52E–03 $^+$ (5.23E–03)	2.99E–03 $^+$ (4.25E–03)
f_{16}	50,000	2.61E–03 (3.85E–03)	2.40E–04 $^+$ (3.71E–04)	1.55E–04 $^+$ (1.51E–04)	2.40E–04 $^+$ (4.87E–04)
f_{17}	50,000	4.71E–03 (1.18E–02)	5.78E–04 $^+$ (9.65E–04)	3.90E–04 $^+$ (7.37E–04)	6 61E–04 $^+$ (1.05E–03)
f_{18}	50,000	9.30E–01 (4.93E+00)	3.44E–03 (7.08E–03)	3.50E–03 (6.32E–03)	2.54E–03 (4.54E–03)
f_{19}	50,000	2.59E–04 (2.37E–04)	2.19E–05 $^+$ (3.62E–05)	1.41E–05 $^+$ (1.66E–05)	2.49E–05 $^+$ (1.97E–05)
f_{20}	50,000	2.83E–02 (5.12E–02)	3.99E–03 $^+$ (2.17E–02)	7.95E–03 $^+$ (3.02E–02)	3.63E–05 $^+$ (3.63E–05)
f_{21}	20,000	5.08E+00 (3.40E+00)	2.70E+00 $^+$ (3.39E+00)	2.95E+00 $^+$ (3.63E+00)	3.60E+00 $^+$ (3.67E+00)
f_{22}	20,000	4.25E+00 (3.44E+00)	3.14E+00 (3.44E+00)	3.91E+00 (3.44E+00)	2.32E+00 (3.44E+00)
f_{23}	20,000	5.01E+00 (3.37E+00)	3.64E+00 $^+$ (3.48E+00)	2.55E+00 $^+$ (3.36E+00)	1.98E+00 $^+$ (3.31E+00)
Win			20 versus 1	20 versus 1	21 versus 1

Table 3.3 Success rate of the basic BBO and Local-BBO on the 13 test functions

f	BBO	RingBBO	SquareBBO	RandBBO
f_1	0	0	0	0
f_2	0	0	0	0
f_3	0	0	0	0
f_4	0	0	0	0
f_5	0	0	0	0
f_6	13.3	100	96.7	96.7
f_7	96.7	83.3	90	93.3
f_8	0	0	0	0
f_9	0	0	0	0
f_{10}	0	0	0	0
f_{11}	0	0	0	0
f_{12}	0	0	0	0
f_{13}	0	0	0	0
f_{14}	13.3	50	56.7	40
f_{15}	0	0	0	0
f_{16}	0	0	0	0
f_{17}	0	0	0	0
f_{18}	0	0	0	0
f_{19}	0	0	0	0
f_{20}	0	0	0	0
f_{21}	0	0	0	0
f_{22}	0	0	0	0
f_{23}	0	0	0	0

an algorithm is less than the accuracy, we consider this running of the algorithm is a "successful" one. Table 3.3 presents the success rates of the algorithms, which show that. As it is shown, the Local-BBO algorithms have significant improvement of the success rate on functions f_6 and f_{14}; on function f_7, the success rate of the basic BBO is slightly better, but there is no significant difference between the basic BBO and the the Local-BBO.

In generally, among the three versions of Local-BBO, there are no statistically significant difference. Roughly speaking, the performance of RandBBO is better than RingBBO and SquareBBO, but RandBBO consumes a bit more computational cost. On most high-dimensional functions, SquareBBO is slightly better than RingBBO in terms of mean bests.

For high-dimensional functions f_1–f_{13}, we present the convergence curves of the four algorithms in Fig. 3.8, from which we can see that the Local-BBO also has much faster convergence speeds on all the functions except f_7, and there are no obvious differences among the convergence speeds of the three Local-BBO versions.

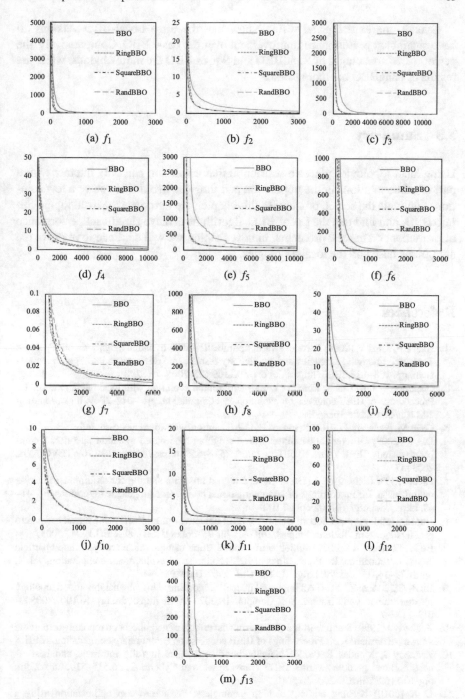

Fig. 3.8 Convergence curves of the four BBO algorithms on the high-dimensional functions

Overall, the experimental results show that the three Local-BBO versions all have significant performance improvement over the basic BBO. Comparatively, the performance advantages of RandBBO and SquareBBO are more obvious, while the results of RingBBO are more stable.

3.5 Summary

Using local topologies places a restriction that a solution can only interact with a part of other solutions in the population and thus can avoid that one or a few solutions dominate the search process. In this chapter, we introduce the local topologies into BBO, and find that the Local-BBO algorithms achieve significant performance improvement on the test problems. In next chapter, we will continue to improve the algorithms based on the local topologies.

References

1. Akat SB, Gazi V (2008) Particle swarm optimization with dynamic neighborhood topology: Three neighborhood strategies and preliminary results. In: IEEE swarm intelligence symposium, pp. 152–159. IEEE. https://doi.org/10.1109/SIS.2008.4668298
2. Chakraborty UK, Das S, Konar A (2006) Differential evolution with local neighborhood. In: Proceedings of IEEE congress on evolutionary computation, pp. 2042–2049. https://doi.org/10.1109/CEC.2006.1688558
3. Clerc M, Kennedy J (2012) Standard PSO 2011. http://www.particleswarm.info
4. Das S (2009) Differential evolution with a neighborhood based mutation operator: a comparative study. IEEE Trans. Evol. Comput. 13:526–553. https://doi.org/10.1109/TEVC.2008.2009457
5. Goldberg D, Richardson J (1987) Genetic algorithms with sharing for multimodal function optimization. In: Proceedings of 2nd international conference on genetic algorithms, pp. 41–49. https://doi.org/10.1002/cplx.20168
6. Horn J, Nafpliotis N, Goldberg DE (1994) Multiobjective optimization using the niched pareto genetic algorithm. Technical report, IlliGAL. https://doi.org/10.1109/ICEC.1994.350037
7. Hu X, Eberhart R (2002) Multiobjective optimization using dynamic neighborhood particle swarm optimization. In: Proceedings of IEEE congress on evolutionary computation, vol. 2, pp. 1677–1681. https://doi.org/10.1109/CEC.2002.1004494
8. Hu Z, Cai X, Fan Z (2014) An improved memetic algorithm using ring neighborhood topology for constrained optimization. Soft Comput. 18:2023–2041. https://doi.org/10.1007/s00500-013-1183-7
9. Kennedy J (1999) Small worlds and mega-minds: effects of neighborhood topology on particle swarm performance. In: Proceedings of IEEE congress on evolutionary computation. IEEE
10. Kennedy J, Mendes R (2006) Neighborhood topologies in fully informed and best-of-neighborhood particle swarms. IEEE Trans. Syst. Man Cybern. C. 36:515–519. https://doi.org/10.1109/TSMCC.2006.875410
11. Li X (2010) Niching without niching parameters: Particle swarm optimization using a ring topology. IEEE Trans. Evol. Comput. 14:150–169. https://doi.org/10.1109/TEVC.2009.2026270

12. Lin L, Wang S-x (2007) Speaker recognition based on adaptive niche hybrid genetic algorithms. Acta Electron. Sin. 35:8–12
13. Miller BL, Shaw MJ (1996) Genetic algorithms with dynamic niche sharing for multimodal function optimization. In: Proceedings of IEEE international conference on evolutionary computation, pp. 786–791
14. Mo S, Zeng J (2009) Performance analysis of the artificial physics optimization algorithm with simple neighborhood topologies. In: Proceedings of international conference on computational intelligence and security (CIS 2009), pp. 155–160. https://doi.org/10.1109/CIS.2009.195
15. Mohais AS, Mendes R, Ward C, Posthoff C (2005) Neighborhood re-structuring in particle swarm optimization, pp. 776–785. Springer, Berlin. https://doi.org/10.1007/11589990-80
16. Omran MGH, Engelbrecht AP, Salman A (2006) Using the ring neighborhood topology with self-adaptive differential evolution. Advances in natual computation, vol 4221. Lecture notes in computer science. Springer, Berlin, pp 976–979. https://doi.org/10.1007/11881070-129
17. Pan QK, Suganthan PN, Liang JJ, Tasgetiren MF (2010) A local-best harmony search algorithm with dynamic subpopulations. Eng. Opt. 42:101–117. https://doi.org/10.1080/03052150903104366
18. Qu BY, Suganthan PN, Liang JJ (2012) Differential evolution with neighborhood mutation for multimodal optimization. IEEE Trans. Evol. Comput. 16:601–614. https://doi.org/10.1109/TEVC.2011.2161873
19. Simon D (2009) Biogeography-based optimization. http://academic.csuohio.edu/simond/bbo/
20. Suganthan PN, Hansen N, Liang JJ, Deb K, Chen Y, Auger A, Tiwari S (2005) Problem definitions and evaluation criteria for the cec 2005 special session on real-parameter optimization. Technical report, KanGAL. http://www.ntu.edu.sg/home/EPNSugan
21. Wei L, Zhao M (2005) A niche hybrid genetic algorithm for global optimization of continuous multimodal functions. Appl. Math. Comput. 160:649–661
22. Lin Y, Hao JM, Ji ZS, Dai YS (2000) A study of genetic algorithm based on isolation niche technique. J. Syst. Eng. 15:86–91
23. Yao X, Liu Y, Lin G (1999) Evolutionary programming made faster. IEEE Trans. Evol. Comput. 3:82–102. https://doi.org/10.1109/4235.771163
24. Zheng Y, Wu X, Ling H, Chen S (2013) A simplified biogeography-based optimization using a ring topology. Advances in swarm intelligence, vol 7928. Lecture notes in computer science. Springer, Berlin, pp 330–337
25. Zheng YJ, Ling HF, Wu XB, Xue JY (2014) Localized biogeography-based optimization. Soft Comput. 18:2323–2334. https://doi.org/10.1007/s00500-013-1209-1

Chapter 4
Ecogeography-Based Optimization: Enhanced by Ecogeographic Barriers and Differentiations

Abstract Ecogeography-based optimization (EBO) is one of the most significant upgrades of BBO, which uses local topologies as the Local-BBO, and further defines two novel migration operators, named local migration and global migration, which borrow ideas from the migration models of ecogeography to enrich information sharing among the solutions. This chapter introduces the EBO algorithm in detail and shows its significant improvement over the basic BBO and Local-BBO in computational experiments.

4.1 Introduction

The use of local topologies can improve BBO by only allowing a special subset of solutions to participate in migration operations, but it does not change the migration operator in essence. Inspired by the migration models of ecogeography, we propose two novel migration operators based on local topologies, which leads to a major upgrade of BBO, named ecogeography-based optimization (EBO) [12].

4.2 Background of Ecogeography

In the basic biogeographical migration model used in BBO, migration can occur between any two habitats, and the migration rates of a habitat only depend on its species richness. Such a model is a great simplification of reality by neglecting ecogeographic isolation among different habitats.

In historical biogeography, Wallace and Darwin [1, 10] considered that species originate in one center of origin, from which some individuals can cross barriers and disperse and then change through natural selection, as illustrated in Fig. 4.1a. This viewpoint is known as the dispersal explanation. In another viewpoint known as the vicariance explanation, the ancestral population was divided into subpopulations by

© Springer Nature Singapore Pte Ltd. and Science Press, Beijing 2019   69
Y. Zheng et al., *Biogeography-Based Optimization: Algorithms and Applications*,
https://doi.org/10.1007/978-981-13-2586-1_4

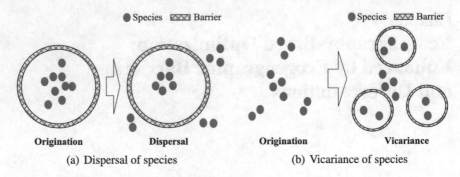

Origination   Dispersal   Origination   Vicariance

(a) Dispersal of species   (b) Vicariance of species

Fig. 4.1 Two main explanations for species distribution

Fig. 4.2 Three kinds of migration routes

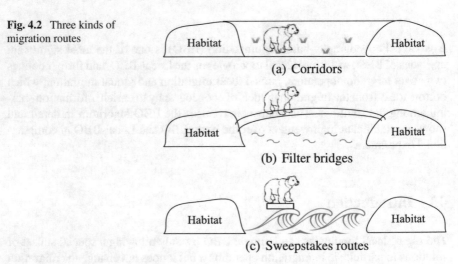

(a) Corridors

(b) Filter bridges

(c) Sweepstakes routes

the development of barriers they cannot cross, and the isolated subpopulations may multiply into different taxa in time [6, 7], as illustrated in Fig. 4.1b. Barrier is a key factor in both the explanations: In the vicariance explanation, the appearance of the barrier causes the disjunction, so the barrier cannot be older than the disjunction. In the dispersal explanation, the barrier is older than the disjunction [5].

In nature, species richness is correlated with many biological, ecological, and geographical factors, and speciation is characterized by the evolution of barriers to genetic exchange between previously interbreeding populations [8]. If two habitats are completely isolated, there is normally no species similarity between them. Otherwise, the migration routes between them can be divided into three classes [4] (which are illustrated in Fig. 4.2):

- Corridors, such as plains and prairies. There is almost no barrier, and the species similarity between the habitats they connect is often very high.

- Filter bridges, such as mountains. Only a part of organisms can pass through, and the similarity between the habitats they connect is inversely proportional to the time required for establishing the connection.
- Sweepstakes routes, such as straits. Only very a few organisms can pass through, and the similarity between the habitats they connect is very low.

When a species migrating to a new habitat, if the ecosystem of the habitat is immature and the species is very suitable to the environment (e.g., there is no natural enemy of the species), then the species will propagate rapidly. On the contrary, if the ecosystem is mature, then the species will take hot competition with the native species, and the result of the competition can be the failure of one side or a new balance between both sides.

4.3 The Ecogeography-Based Optimization Algorithm

The above studies of ecogeography show that migration among habitats depends not only on species richness, but also on many ecological and geographical factors (in particular, ecogeographic barriers). The EBO algorithm takes inspiration from such features to improve the migration model of BBO. First, EBO uses the local topologies described in Sect. 3.2.2 to suppress premature convergence. Moreover, in order to preserve a fast convergence speed, EBO does not completely separate non-neighboring habitats; that is, the migration routes between neighboring habitats are considered as corridors, while the migration routes between non-neighboring habitats are considered as filter bridges and sweepstakes routes. In EBO, migration between neighboring habitats and that between non-neighboring habitats are called local migration and global migration, respectively.

4.3.1 Local Migration and Global Migration

Let H_i be the solution to be immigrated; the local migration on each dimension d is conducted as

$$H_i(d) = H_i(d) + \alpha(H_{nb}(d) - H_i(d)) \tag{4.1}$$

where H_{nb} is a neighboring habitat selected with a probability in proportional to its emigration rate, and α is a coefficient in the range of [0,1], and the item $(H_{nb}(d) - H_i(d))$ represents the "ecological differentiation" between the two habitats.

The global migration is conducted as

$$H_i(d) = \begin{cases} H_{far}(d) + \alpha(H_{nb}(d) - H_i(d)), & f(H_{far}) > f(H_{nb}) \\ H_{nb}(d) + \alpha(H_{far} - H_i(d)), & f(H_{far}) \leq f(H_{nb}) \end{cases} \tag{4.2}$$

where H_{far} is a non-neighboring habitat. In global migration, H_i accepts immigrants from both a neighboring habitat and a non-neighboring habitat: The fitter one between H_{far} and H_{nb} acts as the "primary" immigrant, while the other acts as the "secondary" immigrant that needs to compete with the "original inhabitants" of H_i.

4.3.2   Migration Based on Maturity

Now, we have two new migration operators. But how to determine which operator should be used? EBO uses a parameter η in the range of [0,1] for this purpose: Each migration operation has a probability of η of being a global migration and a probability of $(1-\eta)$ of being a local migration.

During the search process of EBO, we suggest to set η dynamically decrease from an upper limit η_{max} to a lower limit η_{min}:

$$\eta = \eta_{\text{max}} - \frac{t}{t_{\text{max}}}(\eta_{\text{max}} - \eta_{\text{min}}) \tag{4.3}$$

where t is the current iteration number and t_{max} is the total iteration number of the algorithm. This is also inspired by an ecogeographic phenomenon: Global migration is preferred in early stages of evolution, where species are easier to disperse to a wide range of new habitats. With the increase of species richness, the ecological system has more stability or more resistance to invasions. The increase of invasion resistance facilitates divergence, which increases invasion resistance even further. Given enough time, such a feedback loop will lead to a stable connection among habitats. Thereby, in later stages of evolution, the habitats are more likely to accept immigrants from their neighbors. In this sense, η is called the immaturity index representing the immaturity of the ecological system, which is inversely proportional to the invasion resistance of the system. Using this mechanism, EBO prefers more exploration in early stages and emphasizes exploitation in later stages.

4.3.3   The Algorithmic Framework of EBO

Algorithm 4.1 presents the procedure of EBO. Note that both the local migration (4.1) and the global migration (4.2) always produce a component that is different from that of the immigrating habitat and that of the emigrating habitat, which can provide enough diversity to the population. Thus, the random mutation operator of the basic BBO is abandoned in EBO.

Algorithm 4.1: The EBO algorithm

---

1 Randomly initialize a population P of N solutions to the problem;
2 Evaluate the fitness of the solutions and select the best one as H_{best};
3 **while** *the stop criterion is not satisfied* **do**
4   Calculate the migration rates of the solutions, and update η according to Eq. (4.3);
5   **forall the** $H_i \in P$ **do**
6   **forall the** *dimension d of the problem* **do**
7   **if** rand() $< \lambda_i$ **then**
8   Select a neighboring solution H_{nb} with probability $\propto \mu$;
9   **if** rand() $< \eta$ **then**
10   Select a non-neighboring solution H_{far} with probability $\propto \mu$;
11   Perform a global migration according to Eq. (4.2);
12   **else**
13   Perform a local migration according to Eq. (4.1);
14   Evaluate the fitness of the resulting H_i';
15   **if** $f(H_i) > f(H_i')$ **then**
16   Use H_i' to replace H_i in the population;
17   **if** $f(H_i') > f(H_{best})$ **then** set $H_{\text{best}} = H_i'$

18 **return** H_{best};

4.4 Computational Experiments

4.4.1 Experimental Settings

We compare EBO with the basic BBO, the B-BBO [3], the DE method with the DE/rand/1/bin scheme [9], and the hybrid DE/BBO method [2] on a set of 13 test functions, f_1–f_{13}, taken from [11]. Here, we, respectively, conduct experiments on 10-, 30-, and 50-dimensional problems for each test function. For a fair comparison, on each test problem we set MNFE as $5000D$ for all the algorithms, where D is the dimension of the problem. In addition, we record the required NFE (RNFE) of the algorithms to reach the required accuracy, which is set to 10^{-8} for all the functions.

We implement two EBO versions, one with the ring topology and the other with the local random topology, denoted by EBO1 and EBO2, respectively. For EBO2, the neighborhood structure is set such that each habitat has probably $K_N = 2$ neighbors, and it will be reset after every generation where there is no new best solution found.

All the six algorithms use the same population size of 50. We also set $I = E = 1$ for BBO, DE/BBO, and EBO and set $\pi_{\max} = 0.02$ for BBO and DE/BBO. The coefficient α in B-BBO is set to 0.5 which is preferable on most problems [3]. Every algorithm is run for 60 times (with different random seeds) on each problem, and the resulting values are averaged over the 60 runs. The experiments are conducted on a computer of Intel Core i5-2520M processor and 4GB DDR3 memory.

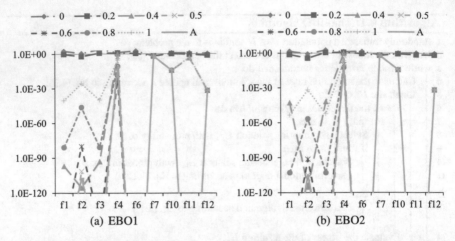

Fig. 4.3 Mean function values obtained by EBO1 and EBO2 with different η values, where 'A' denotes the linearly decreasing value of η

4.4.2   Impact of the Immaturity Index η

First we test the impact of different η values on the performance of EBO. We select nine functions including f_1–f_4, f_6, f_7, and f_{10}–f_{12}, all of which are set to 30 dimension. On each function, we, respectively, set η to seven fixed values including 0, 0.2, 0.4, 0.5, 0.6, 0.8, and 1.0, in addition to a linearly decreasing value with $\eta_{max} = 0.7$ and $\eta_{min} = 0.4$. When η is fixed to 0, the algorithm only uses the local migration operator, and when $\eta = 1$, it only uses the global migration operator.

Figure 4.3a, b presents the mean function error values obtained by EBO1 and EBO2, respectively. For both the algorithms, $\eta = 0$ leads to the worst results on all the functions. In general, the EBO methods with $\eta = 0.4$ or 0.5 perform well on most functions, but on f_4 and f_7 a large η between $0.8 \sim 1.0$ is preferable. However, no single fixed value of η can always be superior to other fixed values. In comparison, when η linearly decreases from 0.7 to 0.4, the overall performance of EBO on the nine functions is more competitive and robust. Thus, we use this dynamic strategy in the following comparative experiments.

4.4.3   Comparison of the 10-D Functions

Table 4.1 presents the comparative results of the six algorithms on the 10-D functions, where "mean" denotes the result function error value averaged over the 60 runs and "std" is the corresponding standard deviation. The best mean error and RNFE values among the six algorithms are shown in bold. From the results, we can see that:

Table 4.1 Experimental results of the six algorithms on the 10-D problems

f	Metric	BBO	B-BBO	DE	DE/BBO	EBO1	EBO2
f_1	mean	1.35E−01	1.06E−03	1.52E−41	3.46E−49	**1.16E−127**	9.25E−124
	std	8.82E−02	1.03E−03	2.60E−41	8.46E−49	3.46E−127	4.93E−123
	RNFE	–	–	11272±398	10667±260	**7143±174**	7193±199
f_2	mean	7.61E−02	1.86E−03	3.32E−25	1.11E−24	**4.62E−68**	5.78E−66
	std	2.34E−02	1.24E−03	2.86E−25	8.66E−25	8.91E−68	1.23E−65
	RNFE	–	–	18212±518	14581±292	**9863±223**	9979±214
f_3	mean	6.27E−01	6.55E−03	1.03E−48	4.18E−41	**2.13E−126**	1.78E−122
	std	4.08E−01	8.12E−03	1.71E−48	7.42E−41	8.76E−126	5.47E−122
	RNFE	–	–	11933±436	11325±271	**7563±172**	7629±201
f_4	mean	8.54E−01	2.18E−01	8.53E−06	8.52E−13	4.93E−30	**1.11E−34**
	std	2.64E−01	7.61E−02	3.95E−05	6.58E−13	3.76E−29	7.64E−34
	RNFE	–	–	28186±7394	35110±865	**18728±496**	18888±526
f_5	mean	4.27E+01	1.92E+02	**1.99E+00**	3.91E+00	2.06E+00	2.32E+00
	std	2.96E+01	8.05E+01	1.33E+00	8.80E−01	2.19E+00	2.22E+00
	RNFE	–	–	–	–	–	–
f_6	mean	**0.00E+00**	**0.00E+00**	**0.00E+00**	**0.00E+00**	**0.00E+00**	**0.00E+00**
	std	0.00E+00	0.00E+00	0.00E+00	0.00E+00	0.00E+00	0.00E+00
	RNFE	27605±8042	10902±2641	4094±316	3834±216	**2534±135**	2566±119

(continued)

Table 4.1 (continued)

f	Metric	BBO	B-BBO	DE	DE/BBO	EBO1	EBO2
f_7	mean	3.84E−02	3.37E−02	8.01E−02	6.67E−02	3.23E−02	**2.66E−02**
	std	2.67E−02	2.22E−02	5.88E−02	4.77E−02	2.23E−02	1.73E−02
	RNFE	–	–	–	–	–	–
f_8	mean	4.00E−01	1.09E−02	2.63E−07	**0.00E+00**	**0.00E+00**	**0.00E+00**
	std	3.40E−01	1.06E−02	2.00E−06	0.00E+00	0.00E+00	0.00E+00
	RNFE	–	–	37353±574	13503±451	**13273±1912.9**	13347±1561
f_9	mean	6.18E−02	4.18E−04	9.80E+00	**0.00E+00**	**0.00E+00**	**0.00E+00**
	std	3.85E−02	7.57E−04	6.58E+00	0.00E+00	0.00E+00	0.00E+00
	RNFE	–	–	–	**16541±771**	19087±1119	19259±1178
f_{10}	mean	2.12E−01	1.12E−01	4.00E−15	4.00E−15	3.70E−15	**3.35E−15**
	std	8.80E−02	2.78E−02	0.00E+00	0.00E+00	9.90E−16	1.39E−15
	RNFE	–	–	18437±535	16743±403	**10118±185**	10300±204
f_{11}	mean	2.45E−01	6.21E−02	2.10E−02	**0.00E+00**	1.32E−10	4.46E−12
	std	8.86E−02	1.14E−01	3.12E−02	0.00E+00	6.56E−10	2.16E−11
	RNFE	–	–	–	30922±3687	30596±8049	**29827±7706**
f_{12}	mean	5.11E−03	4.33E−01	**0.00E+00**	**0.00E+00**	**0.00E+00**	**0.00E+00**
	std	4.82E−03	9.50E−02	0.00E+00	0.00E+00	0.00E+00	0.00E+00
	RNFE	–	–	10783±495	10604±310.6	7968±369	**7951±301**
f_{13}	mean	2.20E−02	3.64E+00	**0.00E+00**	**0.00E+00**	**0.00E+00**	**0.00E+00**
	std	1.66E−02	1.12E+00	0.00E+00	0.00E+00	0.00E+00	0.00E+00
	RNFE	–	–	11827±533	11326±401.9	8456±395	**8425±371**

- All the six algorithms reach the same optimum on function f_6, where EBO1 uses the minimum RNFE.
- DE, DE/BBO, EBO1, and EBO2 reach the same optimum on f_{12} and f_{13}, where EBO2 uses the minimum RNFE on both the functions.
- DE/BBO, EBO1, and EBO2 reach the same optimum on f_8 and f_9, where EBO1 uses the minimum RNFE on f_8, and DE does so on f_9.
- On the remaining eight functions, EBO1 has both the best mean values and the minimum RNFE on f_1–f_3, and EBO2 uniquely achieves the best mean values on f_4, f_7, and f_{10}. DE achieves the best mean value on f_5, where none of the algorithms can reach the required accuracy; DE/BBO does so on f_{11}, but its RNFE is larger than the two EBO methods.

We also conduct paired t tests to identify statistical differences of the result mean values of EBO1/EBO2 from the other comparative algorithms, and present the resulting p-values in Tables 4.2 and 4.3, where † indicates that EBO has statistically significant improvement over the corresponding algorithms at 95% confidence level. The results show that the performance of EBO1 is significantly better than BBO and B-BBO on 11 functions, than DE on 8 functions, and than DE/BBO on 7 functions; EBO2 is significantly better than BBO and B-BBO on 11 functions, than DE 8 functions, and than DE/BBO on 7 functions. By comparing the two EBO versions, EBO1 outperforms EBO2 on two functions, EBO2 outperforms EBO1 on one function, and on the remaining ten functions, there is no significant difference between them. In general, EBO2 has more advantage in the mean values, while EBO1 is more preferable in terms of RNFE.

Table 4.2 Statistical tests between EBO1 and the other algorithms on the 10-D problems

f	BBO	B-BBO	DE	DE/BBO	EBO1
f_1	†4.56E–22	†5.46E–13	†9.87E–04	†7.35E–06	7.46E–02
f_2	†1.17E–49	†1.68E–21	†2.59E–15	†1.49E–17	†2.15E–04
f_3	†2.79E–22	†3.37E–09	†4.56E–06	†1.38E–05	†6.60E–03
f_4	†2.51E–49	†3.59E–44	†4.86E–02	†8.43E–18	8.44E–01
f_5	†3.88E–19	†2.01E–36	5.83E–01	†8.47E–09	2.62E–01
f_6	–	–	–	–	–
f_7	8.92E–02	3.62E–01	†1.37E–06	†2.39E–05	9.38E–01
f_8	†1.19E–15	†4.77E–13	1.60E–01	–	–
f_9	†1.75E–23	†1.95E–05	†2.24E–21	–	–
f_{10}	†1.74E–32	†5.17E–19	†9.00E–03	†9.00E–03	9.46E–01
f_{11}	†8.94E–43	†2.36E–05	†3.92E–07	9.39E–01	9.33E–01
f_{12}	†4.10E–03	†8.03E–65	–	–	–
f_{13}	†1.68E–04	†1.31E–37	–	–	–

Table 4.3 Statistical tests between EBO2 and the other algorithms on the 10-D problems

f	BBO	B-BBO	DE	DE/BBO	EBO2
f_1	[†]4.56E–22	[†]5.46E–13	[†]9.87E–04	[†]7.35E–06	9.25E–01
f_2	[†]1.17E–49	[†]1.68E–21	[†]2.59E–15	[†]1.49E–17	1.00E+00
f_3	[†]2.79E–22	[†]3.37E–09	[†]4.56E–06	[†]1.38E–05	9.93E–01
f_4	[†]2.51E–49	[†]3.59E–44	[†]4.86E–02	[†]8.43E–18	1.56E–01
f_5	[†]5.62E–19	[†]2.26E–36	8.36E–01	[†]4.99E–07	7.38E–01
f_6	–	–	–	–	–
f_7	[†]2.50E–03	[†]2.63E–02	[†]3.04E–08	[†]3.19E–07	6.17E–02
f_8	[†]1.19E–15	[†]4.77E–13	1.60E–01	–	–
f_9	[†]1.75E–23	[†]1.95E–05	[†]2.24E–21	–	–
f_{10}	[†]1.74E–32	[†]5.17E–19	[†]9.00E–03	[†]9.00E–03	5.45E–02
f_{11}	[†]8.94E–43	[†]2.36E–05	[†]3.92E–07	9.44E–01	[†]6.75E–02
f_{12}	[†]4.10E–03	[†]8.03E–65	–	–	–
f_{13}	[†]1.68E–04	[†]1.31E–37	–	–	–

4.4.4   Comparison of the 30-D Functions

Table 4.4 presents the experimental results of the six algorithms on the 30-D functions, and Tables 4.5 and 4.6 present their statistical test results. On this group, the EBO methods also have much performance advantage over the other four, and the advantage is more obvious than that on the 10-D functions. In more details:

- B-BBO, DE/BBO, EBO1, and EBO2 reach the same optimum on function f_6, where EBO1 uses the minimum RNFE.
- DE/BBO, EBO1, and EBO2 reach the same optimum on f_8 and f_{11}, where both EBO1 and EBO2 use the minimum RNFE.
- DE/BBO achieves the best mean value on f_5, where none of the algorithms can reach the required accuracy.
- On the remaining nine test functions, both the mean values and RNFE of the EBO methods are better than the other four; EBO1 achieves the best mean value on seven functions, and EBO2 does so on six functions.

According to the statistical test results, both EBO1 and EBO2 have significant performance improvement over BBO on all the 13 functions and over B-BBO on 12 functions. EBO1 also has significant improvement over DE and DE/BBO on 10 functions and 8 functions, respectively, and EBO2 does so on 11 functions and 8 functions, respectively. By comparing EBO1 and EBO2, they, respectively, outperform the counterpart on three functions, and there is no significant difference between them on the remaining seven functions.

Figure 4.4a–l presents the convergence curves of algorithms on the 30-D functions f_1–f_{12}, respectively (the curves of f_{13} are very similar to those of f_{12}, and the curves of the 10-D and 50-D functions are also similar to the corresponding 30-D functions).

Table 4.4 Experimental results of the six algorithms on the 30-*D* problems

f	Metric	BBO	B-BBO	DE	DE/BBO	EBO1	EBO2
f_1	mean	1.19E+00	1.08E+00	5.66E−51	7.04E−31	**1.46E−187**	3.34E−174
	std	4.65−01	2.42E−01	2.95E−50	1.08E−30	0.00E+00	0.00E+00
	nfe	–	–	33823±1188	37799±571	**16328±294**	17017±285
f_2	mean	3.09E−01	6.10E−02	7.44E−28	1.73E−19	**4.17E−102**	9.82E−95
	std	5.57E−02	1.24E−02	7.10E−28	9.34E−20	6.81E−102	1.14E−94
	nfe	–	–	51844±1187	52597±796	**23720±294**	24728±331
f_3	mean	2.46E+01	2.86E+01	6.85E−49	2.94E−30	**1.07E−183**	3.25E−173
	std	8.75E+00	6.76E+01	4.66E−48	4.62E−30	0.00E+00	0.00E+00
	nfe	–	–	36770±1319	41112±711	**17752±349**	18603±317
f_4	mean	3.05E+00	1.40E+00	8.92E+00	8.03E−04	4.17E−12	**1.63E−13**
	std	5.80E−01	1.57E−01	4.14E+00	2.83E−04	1.12E−11	3.76E−13
	nfe	–	–	–	–	**71563±2154**	73312±1797
f_5	mean	2.60E+02	9.02E+03	2.67E+01	**2.11E+01**	2.24E+01	2.15E+01
	std	2.64E+02	2.49E+03	1.82E+01	4.31E−01	6.25E−01	7.01E−01
	nfe	–	–	–	–	–	–
f_6	mean	8.50E−01	**0.00E+00**	1.83E−01	**0.00E+00**	**0.00E+00**	**0.00E+00**
	std	8.20E−01	0.00E+00	4.31E−01	0.00E+00	0.00E+00	0.00E+00
	nfe	–	–	37506±51725	13997±510.5	**6116±196**	6364±189

(continued)

Table 4.4 (continued)

f	Metric	BBO	B-BBO	DE	DE/BBO	EBO1	EBO2
f_7	mean	1.74E−02	1.37E−02	2.56E−02	2.11E−02	9.80E−03	**7.09E−03**
	std	1.14E−02	8.83E−03	1.70E−02	1.41E−02	7.60E−03	5.02E−03
	nfe	–	–	–	**–**	–	–
f_8	mean	1.92E+00	1.76E+02	2.40E+03	**0.00E+00**	**0.00E+00**	**0.00E+00**
	std	7.83E−01	7.44E−01	1.80E+03	0.00E+00	0.00E+00	0.00E+00
	nfe	–	–	–	67811±5147	**43335±6631**	44783±5673
f_9	mean	3.91E−01	4.32E−02	2.09E+01	2.96E−11	**0.00E+00**	**0.00E+00**
	std	1.58E−01	2.50E−02	1.38E+01	2.18E−10	0.00E+00	0.00E+00
	nfe	–	–	–	124391±10831	90955±9068	103282±10443
f_{10}	mean	2.69E−01	2.70E−02	6.78E−15	6.90E−15	4.00E−15	**4.00E−15**
	std	7.02E−02	7.54E−03	1.86E−15	1.39E−15	0.00E+00	0.00E+00
	nfe	–	–	52873±1778	60562±966	**23125±388**	24348±389
f_{11}	mean	8.16E−01	8.44E−01	2.87E−03	**0.00E+00**	**0.00E+00**	**0.00E+00**
	std	1.44E−01	9.47E−02	8.23E−03	0.00E+00	0.00E+00	0.00E+00
	nfe	–	–	61631±49174	41591±2752	18793±5083	**18485±3252**
f_{12}	mean	8.82E−02	5.65E+00	1.74E−30	1.02E−31	**1.57E−32**	**1.57E−32**
	std	3.07E−02	1.05E+00	2.19E−30	3.38E−31	2.21E−47	2.21E−47
	nfe	–	–	40132±983	36401±3053	**18745±412**	19420±536
f_{13}	mean	4.08E−01	3.91E+00	1.64E−24	4.16E−30	**1.35E−32**	**1.35E−32**
	std	1.35E−01	1.30E+00	1.27E−23	4.77E−30	8.28E−48	8.28E−48
	nfe	–	–	46936±14924	42785±933	**19755±501**	20804±527

Table 4.5 Statistical tests between EBO1 and the other algorithms on the 30-D problems

f	BBO	B-BBO	DE	DE/BBO	EBO1
f_1	†1.03E–39	†4.17E–64	6.99E–02	†8.71E–07	†0.00E+00
f_2	†3.60E–74	†2.22E–68	†2.70E–13	†5.42E–28	†4.16E–10
f_3	†2.20E–43	†1.99E–61	1.28E–01	†1.39E–06	†0.00E+00
f_4	†1.23E–71	†2.03E–97	†4.25E–33	†7.29E–44	9.97E–01
f_5	†9.06E–11	†3.69E–54	†3.59E–02	1.00E+00	1.00E+00
f_6	†4.07E–13	–	†6.58E–04	–	–
f_7	†1.69E–05	†5.50E–03	†7.03E–10	†1.14E–07	9.89E–01
f_8	†5.82E–38	†1.22E–36	†1.59E–18	–	–
f_9	†2.62E–38	†9.91E–26	†6.72E–22	1.47E–01	–
f_{10}	†7.80E–57	†7.02E–54	†1.71E–21	†4.18E–32	5.00E–01
f_{11}	†2.81E–75	†1.35E–97	†3.90E–03	–	–
f_{12}	†2.18E–44	†5.93E–73	†6.96E–09	†2.59E–02	5.00E–01
f_{13}	†1.50E–46	†3.70E–46	1.59E–01	†3.16E–10	5.00E–01

Table 4.6 Statistical tests between EBO2 and the other algorithms on the 30-D problems

f	BBO	B-BBO	DE	DE/BBO	EBO2
f_1	†1.03E–39	†4.17E–64	6.99E–02	†8.71E–07	1.00E+00
f_2	†3.60E–74	†2.22E–68	†2.70E–13	†5.42E–28	1.00E+00
f_3	†2.20E–43	†1.99E 61	†1.28E–01	†1.39E–06	1.00E+00
f_4	†1.23E–71	†2.03E–97	†4.25E–33	†7.29E–44	†3.30E–03
f_5	†7.92E–11	†3.66E–54	†1.47E 02	1.00E+00	†6.15E–12
f_6	†4.07E–13	–	†6.58E–04	–	–
f_7	†1.38E–09	†8.72E–07	†3.04E–13	†1.94E–11	†1.15E–02
f_8	†5.82E–38	†1.22E–36	†1.59E–18	–	–
f_9	†2.62E–38	†9.91e–26	†6.72E–22	1.47E–01	–
f_{10}	†7.80E 57	†7.02E–54	†1.71E–21	†4.18E–32	5.00E–01
f_{11}	†2.81E–75	†1.35E–97	†3.90E–03	–	–
f_{12}	†2.18E–44	†5.93E–73	†6.96E–09	†2.59E–02	5.00E–01
f_{13}	†1.50E–46	†3.70E–46	1.59E–01	†3.16E–10	5.00E–01

As we can see, the overall convergence speed of EBO is much better than the other algorithms on almost all of the problems. On some problems (such as f_5–f_8), BBO and B-BBO converge fast at the very early stage, but their curves soon become flat. DE/BBO typically converges faster than BBO and slower than DE, but it often converges longer and thus reaches better results than DE. In comparison, the curves of EBO fall not only fast but also deep, which demonstrates that they achieve a much better balance between exploration and exploitation.

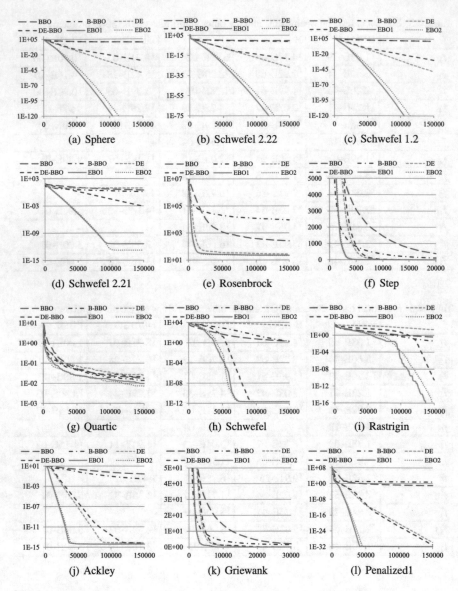

Fig. 4.4 Convergence curves of the comparative algorithms on the 30-D problems

4.4.5 Comparison of the 50-D Functions

Table 4.7 presents the experimental results of the six algorithms on the 50-D functions, and Tables 4.8 and 4.9 present their statistical test results. On this group:

- DE/BBO, EBO1, and EBO2 reach the same optimum on function f_6, where EBO1 uses the minimum RNFE.
- BBO achieves the best mean value on f_8, where none of the algorithms can reach the required accuracy.
- On the remaining 11 test functions, the EBO methods outperform all the other four. Individually, EBO1 has the best mean values on two functions, EBO2 does so on six functions, and the two methods both achieve the same best results on three functions.

According to the statistical test results, EBO1 has significant performance improvement over BBO, B-BBO, DE, and DE/BBO on 11, 12, 11, and 9 functions, respectively, and EBO2 does so on 11, 12, 11, and 10 functions, respectively. EBO1 outperforms EBO2 on two functions, while EBO2 outperforms EBO1 on five functions.

4.4.6 Discussion

In summary, the EBO methods exhibit great performance advantages over the other algorithms on the benchmark problems, and the advantages become more obvious with the increase of problem dimension. The basic BBO and B-BBO sometimes converge fast at the very early stage, but they are easy to be trapped in local optima because of poor exploration abilities. DE/BBO combines the DE's ability in exploration and BBO's ability in exploitation and thus keeps a fast convergency speed longer than BBO and exploits more precise optima than DE. EBO achieves a much better balance between exploration and exploitation due to the two new migration operators, and they are more capable of escaping from local optima by using the immaturity index η.

Comparing EBO1 with EBO2, the former generally converges faster, while the latter achieves better results on more test functions. This indicates the local random topology can produce results slightly better than the ring topology, but the ring topology is easier to implement and consumes less computational time. The random topology exhibits more performance advantage on higher dimensional problems. In general, we prefer to use the random topology in EBO for most unknown global optimization problems and favor the use of the ring topology for urgent problems.

Table 4.7 Experimental results of the six algorithms on the 50-D problems

f	Metric	BBO	B-BBO	DE	DE/BBO	EBO1	EBO2
f_1	mean	2.67E+00	1.13E+01	2.97E−46	2.38E−27	**6.15E−184**	3.86E−182
	std	7.14E−01	2.12E+00	1.83E−45	1.65E−27	0.00E+00	0.00E+00
	nfe	–	–	60775±4132	77685±1177	**24793±406**	26615±423
f_2	mean	6.16E−01	3.79E−01	8.81E−29	3.14E−17	**5.47E−110**	1.57E−104
	std	7.75E−02	5.35E−02	2.80E−28	1.63E−17	2.68E−109	3.51E−104
	nfe	–	–	81840±2472	107976±1524	**36543±366**	39006±481
f_3	mean	1.11E+02	4.20E+02	1.46E−44	1.53E−26	3.69E−178	**2.47E−179**
	std	3.06E+01	7.38E+01	7.95E−44	1.69E−26	0.00E+00	0.00E+00
	nfe	–	–	66516±3714	85038±1255	**27198±439**	29228±392
f_4	mean	4.90E+00	1.16E+00	1.96E+01	3.75E−01	1.06E−05	**3.23E−07**
	std	5.84E−01	1.30E−01	4.71E+00	7.78E−02	1.61E−05	6.00E−07
	nfe	–	–	–	–	–	215916±45246
f_5	mean	4.43E+02	1.49E+03	7.10E+01	4.10E+01	4.00E+01	**3.88E+01**
	std	2.32E+02	3.23E+02	3.56E+01	9.73E+00	7.11E−01	5.39E−01
	nfe	–	–	–	–	–	–
f_6	mean	2.13E+00	1.44E+01	5.18E+00	**0.00E+00**	**0.00E+00**	**0.00E+00**
	std	1.40E+00	3.30E+00	1.16E+01	0.00E+00	0.00E+00	0.00E+00
	nfe	–	–	–	29273±1060	**9358±277**	10138±364

(continued)

Table 4.7 (continued)

f	Metric	BBO	B-BBO	DE	DE/BBO	EBO1	EBO2
f_7	mean	1.38E−02	1.49E−02	0.015186991	1.33E−02	4.94E−03	**3.40E−03**
	std	7.59E−03	6.75E−03	0.011797576	9.58E−03	3.98E−03	2.70E−03
	nfe	–	–	–	–	–	–
f_8	mean	**4.00E+00**	7.60E+01	4.21E+03	1.01E+02	4.25E+02	3.27E+02
	std	1.16E+00	4.42E+01	2.69E+03	1.21E+02	1.75E+02	1.68E+02
	nfe	–	–	–	–	–	–
f_9	mean	8.05E−01	9.65E−01	3.12E+01	1.43E+01	7.30E−01	**6.09E−01**
	std	2.69E−01	5.14E−01	7.83E+00	1.19E+01	7.74E−01	7.08E−01
	nfe	–	–	–	–	–	–
f_{10}	mean	2.80E−01	7.66E−02	9.74E−02	1.07E−14	6.96E−15	**6.72E−15**
	std	5.23E−02	1.25E−02	2.97E−01	2.78E−15	1.34E−15	1.52E−15
	nfe	–	–	–	126620±1889	**34884±480**	37540±498
f_{11}	mean	9.77E−01	1.10E+00	3.32E−03	1.85E−18	**0.00E+00**	**0.00E+00**
	std	7.57E−02	2.00E−02	6.70E−03	1.43E−17	0.00E+00	0.00E+00
	nfe	–	–	104991±80730	81130±2079	**25186.7±3690.8**	27703.3±4315.0
f_{12}	mean	2.15E−01	1.51E+00	1.28E−02	1.46E−26	**1.57E−32**	**1.57E−32**
	std	4.14E−02	3.41E−01	3.88E−02	2.21E−26	6.25E−35	6.25E−35
	nfe	–	–	105568±50594	84062±1604	**28798±552**	31018±695
f_{13}	mean	1.31E+00	2.82E+01	6.07E−01	6.86E−26	**1.35E−32**	**1.35E−32**
	std	3.32E−01	6.69E+00	1.42E+00	8.28E−26	8.28E−48	8.28E−48
	nfe	–	–	213305±43033	90758±2169	**30763±753**	33022±858

Table 4.8 Statistical tests between EBO2 and the other algorithms on the 50-D problems

f	BBO	B-BBO	DE	DE/BBO	EBO1
f_1	†9.82E–56	†1.98E–72	1.06E–01	†1.55E–20	†0.00E+00
f_2	†7.42E–92	†4.34E–86	†8.20E–03	†3.22E–29	†3.76E–04
f_3	†2.73E–54	†1.75E–75	7.93E–02	†9.40E–11	1.00E+00
f_4	†1.50E–94	†1.36E–97	†1.02E–60	†1.86E–67	1.00E+00
f_5	†5.90E–26	†4.19E–64	†2.98E–10	2.12E–01	1.00E+00
f_6	†4.28E–22	†7.77E–63	†3.60E–04	–	–
f_7	†4.01E–13	†2.54E–17	†1.84E–09	†3.66E–09	9.93E–01
f_8	1.00E+00	1.00E+00	†9.03E–20	1.00E+00	9.99E–01
f_9	2.38E–01	†2.62E–02	†2.68E–57	†5.73E–15	8.12E–01
f_{10}	†1.73E–72	†7.01E–79	†6.10E–03	†3.22E–16	8.17E–01
f_{11}	†3.45E–116	†3.58E–190	†9.92E–05	1.60E–01	–
f_{12}	†5.11E–71	†1.31E–63	†5.90E–03	†6.20E–07	–
f_{13}	†4.87E–58	†3.76E–61	†6.10E–04	†1.56E–09	–

Table 4.9 Statistical tests between EBO2 and the other algorithms on the 50-D problems

f	BBO	B-BBO	DE	DE/BBO	EBO2
f_1	†9.82E–56	†1.98E–72	1.06E–01	†1.55E–20	1.00E+00
f_2	†7.42E–92	†4.34E–86	†8.20E–03	†3.22E–29	1.00E+00
f_3	†2.73E–54	†1.75E–75	7.93E–02	†9.40E–11	†0.00E+00
f_4	†1.50E–94	†1.36E–97	†1.02E–60	†1.85E–67	†1.26E–06
f_5	†4.77E–26	†3.84E–64	†8.07E–11	†4.09E–02	†1.22E–18
f_6	†4.28E–22	†7.77E–63	†3.60E–04	–	–
f_7	†8.49E–18	†4.94E–23	†5.17E–12	†2.37E–12	†7.30E–03
f_8	1.00E+00	1.00E+00	†1.92E–20	1.00E+00	†1.10E–03
f_9	2.36E–01	†1.00E–03	†1.63E–57	†3.69E–15	1.88E–01
f_{10}	†1.73E–72	†7.01E–79	†6.10E–03	†5.16E–17	1.83E–01
f_{11}	†3.45E–116	†3.58E–190	†9.92E–05	1.60E–01	–
f_{12}	†5.11E–71	†1.31E–63	†5.90E–03	†6.20E–07	–
f_{13}	†4.87E–58	†3.76E–61	†6.10E–04	†1.56E–09	–

4.5 Summary

This chapter presents EBO, a novel BBO variant which not only employs local topologies but also defines two new migration operators to improve the exploration ability without harming the exploitation ability of BBO. Computational experiments demonstrate that the EBO is a very competitive method for global optimization. EBO has already been applied to a variety of real engineering optimization problems, which will be presented in the subsequent chapters.

References

1. Darwin C (1859) On the origin of species by means of natural selection, or the preservation of favoured races in the struggle for life. Murray, London. http://www.talkorigins.org/faqs/origin.html
2. Gong W, Cai Z, Ling CX (2010) DE/BBO: a hybrid differential evolution with biogeography-based optimization for global numerical optimization. Soft Comput 15:645–665. https://doi.org/10.1007/s00500-010-0591-1
3. Ma H, Simon D (2011) Analysis of migration models of biogeography-based optimization using markov theory. Eng Appl Artif Intell 24:1052–1060. https://doi.org/10.1016/j.engappai.2011.04.012
4. McKenna M (1973) Sweepstakes, filters, corridors, Noahs arks, and beached viking funeral ships in Palaeogeography. In: Implications of continental drift to the earth sciences. Academic Press, London, pp 295–308
5. Morrone JJ, Crisci JV (1995) Historical biogeography: introduction to methods. Annu Rev Ecol Syst 26:373–401. https://doi.org/10.1146/annurev.es.26.110195.002105
6. Nelson G, Platnick NI (1980) A vicariance approach to historical biogeography. Bioscience 30:339–343. https://doi.org/10.2307/1307856
7. Nelson G, Platnick N (1981) Systematics and biogeography: cladistics and vicariance. Columbia University Press, New York
8. Rieseberg L, Willis J (2007) Plant speciation. Science 317:910–914
9. Storn R, Price K (1997) Differential evolution - a simple and efficient heuristic for global optimization over continuous spaces. J Global Optim 11:341–359. https://doi.org/10.1023/A:1008202821328
10. Wallace CC, Pandolfi JM, Young A, Wolstenholme J (1991) Indo-pacific coral biogeography: a case study from the acropora selago group. Aust Syst Bot 4:199–210. https://doi.org/10.1071/SB9910199
11. Yao X, Liu Y, Lin G (1999) Evolutionary programming made faster. IEEE Trans Evol Comput 3:82–102. https://doi.org/10.1109/4235.771163
12. Zheng YJ, Ling HF, Xue JY (2014) Ecogeography-based optimization: enhancing biogeography-based optimization with ecogeographic barriers and differentiations. Comput Oper Res 50:115–127. https://doi.org/10.1016/j.cor.2014.04.013

Chapter 5
Hybrid Biogeography-Based Optimization Algorithms

5.1 Introduction

In the previous chapters, we introduce how to improve the basic BBO algorithm by using local topologies and developing new migration operators. Besides improving the intrinsic structure and operators of the algorithm, another way to improve the algorithm is combining it with other heuristic algorithms. Based on its distinctive migration mechanisms, BBO has a good local exploitation ability [10], but its global exploration ability is relatively poor. Thus, those hybrid BBO algorithms often introduce effective global exploration mechanisms of other heuristic algorithms, so as to better balance the global and local search. This chapter describes some typical hybrid BBO algorithms.

5.2 Hybridization with Differential Evolution

The differential mutation schemes make DE very efficient for exploring large search space to quickly locate those regions containing the optimal or near-optimal solutions [11], but its capability in exploiting a small region is limited. Thus, DE mutation and BBO migration can complement each other, and there are several different approaches for hybridizing BBO with DE.

5.2.1 The DE/BBO Algorithm

Gong et al. [3] propose a hybrid DE with BBO algorithm, named DE/BBO algorithm. The central of DE/BBO is a hybrid migration operator integrated with DE mutation, the procedure of which is shown in Algorithm 5.1, where N is the population size, D is the problem dimension, c_r is the crossover probability, and F is the scaling factor

© Springer Nature Singapore Pte Ltd. and Science Press, Beijing 2019
Y. Zheng et al., *Biogeography-Based Optimization: Algorithms and Applications*,
https://doi.org/10.1007/978-981-13-2586-1_5

in DE. As we can see, the hybrid migration operator produces an offspring U_i whose components are possibly from three sources:

(1) From its parent H_i with a probability $(1 - \lambda_i)$, which makes better solutions be less destroyed.
(2) From the DE mutant with a probability $\lambda_i(c_r + 1/D)$, which puts emphasis on exploring new areas in the search space and helps to improve the solution diversity.
(3) From the emigrant of other solutions with a probability $\lambda_i(1 - c_r - 1/D)$, which enables good solutions to exploit the area nearby and makes poor solutions accept more features from the good solutions.

Algorithm 5.1: The hybrid migration operation (on each solution H_i) in DE/BBO

---

1 Create an empty solution vector U_i of length D;
2 Generate three distinct indices $r_1, r_2, r_3 \in [1, N]$;
3 Generate a random integer $d_r \in [1, D]$;
4 **for** $d = 1$ *to* D **do**
5   **if** rand$() < \lambda_i$ **then**
6   **if** rand$() < c_r$ *or* $d = d_r$ **then**
7              $U_i(d) = H_{r_1}(d) + F \cdot (H_{r_2}(d) - H_{r_3}(d))$;   // DE mutation
8   **else**
9   Select another solution H_j with a probability $\propto \mu_j$;   // BBO migration
10             $U_i(d) = H_j(d)$;
11   **else** $U_i(d) = H_i(d)$;
12 **return** U_i;

DE/BBO replaces the original mutation operator with the above hybrid migration operator in DE. Algorithm 5.2 presents the framework of DE/BBO. In comparison with the basic BBO, DE/BBO only incurs a little additional computational overhead of calculating migration rates. By using the original DE selection operator, DE/BBO also implies an elite strategy that always keeps the current best solution in the population. Thereby, DE/BBO structure combines desired advantages of individual DE and BBO, without increasing the complexity of algorithmic structure too much.

DE/BBO is compared with DE and BBO on the set of benchmark functions f_1-f_{23} (see Table 3.1 for detailed information) taken from [16]. The parameters of DE/BBO are set as $N = 100$, $I = E = 1$, $c_r = 0.9$, and F be a random value in $[0.1, 1]$. The parameters of DE and BBO are set as in the literature.

Each algorithm is run on each function for 50 times independently. In Table 5.1, columns 3–8 present the mean best error values obtained by the algorithms on the functions, where the best result among the three algorithms on every function is marked in bold, and the last two columns present the results of paired t-test on DE/BBO versus DE (denoted as "1 versus 2") and DE/BBO versus BBO (denoted

---

Algorithm 5.2: The DE/BBO algorithm

---

1 Initialize a population of solutions;
2 Evaluate fitness of the solutions;
3 **while** *the stop criterion is not satisfied* **do**
4   Calculate the migration rates of the solutions;
5   **forall the** *solution H_i in the population* **do**
6   Generate a new solution U_i according to Algorithm 5.1;
7   Calculate the fitness of U_i;
8   **if** U_i *is better than* H_i **then** Replace H_i with U_i in the population

9 **return** *the best solution found so far;*

---

as "1 versus 3"), where a superscript [†] means that the DE/BBO is significantly better than the corresponding other algorithm and * means the opposite.

Compared with DE, DE/BBO is significantly better on eight functions, while DE is significantly better DE/BBO only on f_3 and f_5. There are no statistically significant differences between DE and DE/BBO on the other 13 functions. In particular, on multimodal functions f_8-f_{13}, DE/BBO can always achieve the required accuracy of 10^{-8}, but DE can only does so on f_{10}, i.e., on the remaining five functions DE is trapped in local optima. This indicates than DE/BBO is more capable of escaping from poor local optima than DE.

Compared with BBO, DE/BBO is significantly better on all functions. In terms of convergence speed, BBO converges faster at beginning, while DE/BBO evolves steadily to improve its solutions during the whole evolutionary process, and finally achieves much higher solution accuracy than BBO. This shows that the hybrid migration of DE/BBO can balance the global and local search much better than the basic BBO.

5.2.2 Local-DE/BBO

Chapter 3 introduces Local-BBO [21] that improves the basic BBO by using local topologies. The work in [21] also equips DE/BBO with the local topologies to further improve the algorithm, which results in three versions of Local-DE/BBO algorithms, named RingDB, SquareDB, and RandDB, which equips DE/BBO with the ring topology, the square topology, and the local random topology, respectively.

The BBO migration operators based on these three topologies have been introduced in Sect. 3.3. If we modify "Select another solution H_j;" to "Select another solution H_j from the neighborhood" in line 9 of Algorithm 5.1, then Algorithm 5.2 becomes the corresponding Local-DE/BBO algorithm. For the convenience of readers, Algorithm 5.3 presents the procedure of hybrid mutation under the ring topology.

To verify the effects of the local topologies on the performance of DE/BBO, similarly, we use the same 23 benchmark functions as in the DE/BBO experiments

Table 5.1 Experimental results of DE/BBO, DE, and BBO algorithms on the test problems [3]

f	MNFE	DE/BBO		DE		BBO		t-test	
		Mean	Std	Mean	Std	Mean	Std	1 versus 2	1 versus 3
f_1	150,000	**8.66E−28**	5.21E−28	1.10E−19	1.34E−19	8.86E−01	3.26E−01	−5.81†	−19.21†
f_2	200,000	**0.00E+00**	0.00E+00	1.66E−15	8.87E−16	2.42E−01	4.58E−02	−13.24†	−37.36†
f_3	500,000	2.26E−03	1.58E−03	**8.19E−12**	1.65E−11	4.16E+02	2.02E+02	10.10*	−14.58†
f_4	500,000	**1.89E−15**	8.85E−16	7.83E+00	3.78E−01	7.76E−01	1.72E−01	−14.65†	−31.96†
f_5	500,000	1.90E+01	7.52E+00	**8.41E−01**	1.53E+00	9.14E+01	3.78E+01	16.73*	−13.28†
f_6	150,000	**0.00E+00**	0.00E+00	**0.00E+00**	0.00E+00	2.80E−01	5.36E−01	0	−3.69†
f_7	300,000	**3.44E−03**	8.27E−04	3.49E−03	9.60E−04	1.90E−02	7.29E−03	−0.29	−14.96†
f_8	300,000	**0.00E+00**	0.00E+00	4.28E+02	4.69E+02	5.09E−01	1.65E−01	−6.45†	−21.78†
f_9	300,000	**0.00E+00**	0.00E+00	1.14E+01	7.57E+00	8.50E−02	3.42E−02	−10.61†	−17.61†
f_{10}	150,000	**1.07E−14**	1.90E−15	6.73E−11	2.86E−11	3.48E−01	7.06E−02	−16.66†	−34.81†
f_{11}	200,000	**0.00E+00**	0.00E+00	1.23E−03	3.16E−03	4.82E−01	1.27E−01	−2.76†	−26.93†
f_{12}	150,000	**7.16E−29**	6.30E−29	2.07E−03	1.47E−02	5.29E−03	5.21E−03	−1.00	−7.18†
f_{13}	150,000	**9.81E−27**	7.10E−27	2.75E−13	5.09E−01	1.42E−01	5.14E−02	−1.00	−19.50†
f_{14}	10,000	**0.00E+00**	0.00E+00	**4.94E−19**	1.55E−12	8.85E−06	2.74E−05	−1.26	−2.28†
f_{15}	40,000	3.84E−12	2.70E−11	**1.98E−13**	5.20E−19	5.92E−04	2.68E−04	1.01	−15.65†
f_{16}	10,000	1.15E−12	6.39E−12	7.32E−10	4.12E−13	6.75E−04	1.09E−03	1.05	−4.37†
f_{17}	10,000	**2.92E−10**	1.38E−09	9.14E−13	2.21E−09	4.39E−04	4.26E−04	−1.19	−7.30†
f_{18}	10,000	9.15E−13	6.51E−15	1.36E−14	5.01E−15	7.86E−03	9.57E−03	0.70	−5.81†
f_{19}	10,000	**0.00E+00**	0.00E+00	1.36E−14	6.40E−15	2.51E−04	2.62E−04	−15.05†	−6.76†
f_{20}	20,000	**0.00E+00**	0.00E+00	4.76E−03	2.35E−02	1.46E−02	3.90E−02	−1.43	−2.64†
f_{21}	10,000	3.59E−03	1.44E−02	**6.83E−06**	1.26E−05	5.18E+00	3.34E+00	1.76	−10.95†
f_{22}	10,000	**3.14E−07**	1.36E−06	7.26E−06	4.53E−05	3.67E+00	3.40E+00	−1.08	−7.63†
f_{23}	10,000	**2.50E−08**	5.37E−08	3.70E−06	1.75E−05	2.73E+00	3.29E+00	−1.49	−5.87†

Algorithm 5.3: The hybrid migration operation in RingDB, the Local-DE/BBO based on the ring topology

1 Create an empty solution vector U_i of length D;
2 Generate three distinct indices $r_1, r_2, r_3 \in [1, N]$;
3 Generate a random integer $d_r \in [1, D]$;
4 **for** $d = 1$ *to* D **do**
5   **if** rand$() < \lambda_i$ **then**
6   **if** rand$() < c_r$ *or* $d = d_r$ **then**
7              $U_i(d) = H_{r_1}(d) + F \cdots (H_{r_2}(d) - H_{r_3}(d))$;   // DE mutation
8   **else**
9   Let $i_1 = (i - 1)\%N, i_2 = (i + 1)\%N$;
10   **if** rand$() < \mu_{i1}/(\mu_{i_1} + \mu_{i_2})$ **then** $U_i(d) = H_{i1}(d)$;
11   **else** $U_i(d) = H_{i2}(d)$;
12   **else** $U_i(d) = H_i(d)$

---

[3] to compare the RingDB, SquareDB, RandDB, DE, and DE/BBO algorithms. For the first three algorithms, we set $F = 0.5, c_r = 0.9$, and other parameters as the same as the Local-BBO in Chap. 3. The parameters of DE and DE/BBO are set as same as the experiments in Sect. 5.2.1.

On every function, every algorithm is run for 60 times with the same terminal condition that the algorithm's NFE reaches the MNFE. Table 5.2 presents the experimental results. In columns 3–7, the upper part of each cell is the mean best error value and the lower part is the corresponding standard deviation. As we can see, the mean best values of three Local-DE/BBO algorithms are much better than those of the original DE/BBO with the global topology on functions f_1, f_2, f_3, f_7, and f_{10}, while the original DE/BBO has slight performance advantage only on two unimodal functions f_4 and f_5. In addition, because the hybrid migration in DE/BBO is very effective for multimodal functions, the four DE/BBO algorithms find the optimal solutions (or very near-optimal solutions whose errors can be negligible) on f_8, f_9, f_{11}, f_{12}, and f_{13}.

We also perform rank-sum tests on the experimental results. In column $5 \sim 7$ of Table 5.2, a superscript "•" means that the Local-DE/BBO has statistically significant improvement over DE/BBO and "○" vice versa; "+" means that the Local-DE/BBO has statistically significant improvement over DE and "-" vice versa. The last row presents the statistical information, the first line is the number of functions on which Local-DE/BBO has significant improvement over DE/BBO versus the number of functions on which Local-DE/BBO does not, and the second line is the number of functions on which Local-DE/BBO has significant improvement over DE versus that Local-DE/BBO does not. The test results can be described as follows:

- The three Local-DE/BBO algorithms have significant improvement over DE/BBO on high-dimensional functions f_1, f_2, f_7, and f_{10}.
- DE/BBO has significant advantage over RingDB and RandDB on f_4, and over SquareDB and RandDB on f_5.

Table 5.2 Experimental results of DE, DE/BBO, and Local-DE/BBO algorithms in the 23 test problems [21]

f	MNFE	DE	DE/BBO	RingDB	SquareDB	RandDB
f_1	150,000	2.69E–51 (7.28E–51)	5.48E–57$^+$ (6.70E–57)	2.96E–83$^{•+}$ (4.40E–83)	1.92E–71$^{•+}$ (1.49E–70)	5.92E–70$^{•+}$ (4.01E–69)
f_2	200,000	8.38E–28 (1.20E–27)	5.27E–33$^+$ (2.63E–33)	3.04E–50$^{•+}$ (2.56E–50)	4.79E–49$^{•+}$ (2.63E–48)	6.59E–47$^{•+}$ (3.59E–46)
f_3	500,000	1.09E–178 (0.00E+00)	2.93E–198 (0.00E+00)	1.79E–285 (0.00E+00)	4.41E–225 (0.00E+00)	5.81E–216 (0.00E+00)
f_4	500,000	9.89E+00 (4.33E+00)	1.84E–09$^+$ (1.01E–08)	2.32E–05° (7.84E–05)	3.47E–05 (1.36E–04)	6.63E–05° (2.40E–04)
f_5	500,000	1.46E+01 (4.13E+00)	2.57E+02$^-$ (6.17E–01)	2.60E+02$^-$ (1.37E+01)	2.63E+02$^{\circ-}$ (6.84E–01)	2.62E+02$^{\circ-}$ (1.41E+01)
f_6	150,000	3.33E–02 (1.83E–01)	0.00E+00 (0.00E+00)	0.00E+00 (0.00E+00)	0.00E+00 (0.00E+00)	0.00E+00 (0.00E+00)
f_7	300,000	7.22E–03 (5.49E–03)	5.66E–03 (4.03E–03)	1.77E–03$^{•+}$ (1.42E–03)	1.81E–03$^{•+}$ (1.11E–03)	1.60E–03$^{•+}$ (9.26E–04)
f_8	300,000	4.99E+02 (3.0E+02)	0.00E+00$^+$ (0.00E+00)	0.00E+00$^+$ (0.00E+00)	0.00E+00$^+$ (0.00E+00)	0.00E+00$^+$ (0.00E+00)
f_9	300,000	1.32E+01 (3.62E+00)	0.00E+00$^+$ (0.00E+00)	0.00E+00$^+$ (0.00E+00)	0.00E+00$^+$ (0.00E+00)	0.00E+00$^+$ (0.00E+00)
f_{10}	150,000	7.19E–15 (1.95E–15)	5.30E–15$^+$ (1.74E–15)	4.47E–15$^{•+}$ (1.23E–15)	4.23E–15$^{•+}$ (9.01E–16)	4.59E–15$^{•+}$ (1.35E–15)
f_{11}	200,000	2.96E–03 (5.31E–03)	0.00E+00$^+$ (0.00E+00)	0.00E+00$^+$ (0.00E+00)	0.00E+00$^+$ (0.00E+00)	0.00E+00$^+$ (0.00E+00)
f_{12}	150,000	2.00E–32 (1.02E–32)	1.57E–32 (0.00E+00)	1.57E–32 (0.00E+00)	1.57E–32 (0.00E+00)	1.57E–32 (0.00E+00)
f_{13}	150,000	3.30E–18 (1.81E–17)	1.35E–32 (0.00E+00)	1.35E–32 (0.00E+00)	1.35E–32 (0.00E+00)	1.35E–32 (0.00E+00)
f_{14}	10,000	0.00E+00 (0.00E+00)	0.00E+00 (0.00E+00)	0.00E+00 (0.00E+00)	0.00E+00 (0.00E+00)	0.00E+00 (0.00E+00)
f_{15}	40,000	5.53E–19 (9.16E–20)	7.39E–05$^-$ (8.76E–05)	1.35E–04$^-$ (1.31E–04)	1.27E–04$^-$ (1.26E–04)	1.33E–04$^-$ (1.87E–04)
f_{16}	10,000	1.22E–13 0.00E+00	1.22E–13 0.00E+00	1.22E–13 0.00E+00	1.22E–13 0.00E+00	1.22E–13 0.00E+00
f_{17}	10,000	1.74E–16 0.00E+00	1.74E–16 0.00E+00	1.74E–16 0.00E+00	1.74E–16 0.00E+00	1.74E–16 0.00E+00
f_{18}	10,000	9.59E–14 0.00E+00	9.59E–14 0.00E+00	9.59E–14 0.00E+00	9.59E–14 0.00E+00	9.59E–14 0.00E+00
f_{19}	10,000	0.00E+00 (0.00E+00)	0.00E+00 (0.00E+00)	0.00E+00 (0.00E+00)	0.00E+00 (0.00E+00)	0.00E+00 (0.00E+00)
f_{20}	20,000	0.00E+00 (0.00E+00)	0.00E+00 (0.00E+00)	0.00E+00 (0.00E+00)	0.00E+00 (0.00E+00)	0.00E+00 (0.00E+00)
f_{21}	10,000	1.68E–01 (9.22E–01)	2.49E–01 (1.36E+00)	4.59E–01 (1.75E+00)	2.56E–01 (1.36E+00)	1.20E–01 (5.17E–01)
f_{22}	10,000	1.81E–10 (8.95E–16)	1.81E–10 (6.73E–16)	1.81E–10 (7.99E–16)	1.81E–10 (6.73E–16)	1.81E–10 (8.95E–16)
f_{23}	10,000	2.00E–15 (5.42E–16)	2.00E–15 (5.42E–16)	2.13E–15 (7.23E–16)	1.78E–15 (0.00E+00)	2.07E–15 (6.73E–16)
Win			7 versus 2	4 versus 1 7 versus 2	4 versus 1 7 versus 2	4 versus 2 7 versus 2

Table 5.3 Success rates of DE, DE/BBO, and Local-DE/BBO algorithms on 23 test problems [21]

f	DE	DE/BBO	RingDB	SquareDB	RandDB
f_1	100	100	100	100	100
f_2	100	100	100	100	100
f_3	100	100	100	100	100
f_4	0	96.7	80	56.7	63.3
f_5	0	0	0	0	0
f_6	96.7	100	100	100	100
f_7	70	83.3	100	100	100
f_8	10	100	100	100	100
f_9	0	100	100	100	100
f_{10}	100	100	100	100	100
f_{11}	100	100	100	100	100
f_{12}	100	100	100	100	100
f_{13}	100	100	100	100	100
f_{14}	100	100	100	100	100
f_{15}	100	20	3.33	13.3	13.3
f_{16}	100	100	100	100	100
f_{17}	100	100	100	100	100
f_{18}	100	100	100	100	100
f_{19}	100	100	100	100	100
f_{20}	100	100	100	100	100
f_{21}	96.7	96.7	90	93.3	93.3
f_{22}	100	100	100	100	100
f_{23}	100	100	100	100	100

- The three Local-DE/BBO algorithms have significant advantages over DE on high-dimensional functions f_1, f_2, and f_7–f_{10}.
- DE has significant advantage over Local-DE/BBO only on high-dimensional function f_5 and low-dimensional function f_{15}.
- On the noisy function f_7, there is no statistically significant differences between DE and DE/BBO, but the three Local-DE/BBO algorithms have significant improvement over DE.

Table 5.3 presents the success rates of the algorithm. As we can see, the success rates of three Local-DB/BBO algorithms are much higher than DE on f_4, f_6, f_7, f_8, and f_9 and higher than the original DE/BBO on f_7. However, the success rates of DE are higher on f_{15} and f_{21}, while the success rates of DE/BBO are higher on f_4 and f_{21}.

Figure 5.1 presents the convergence curves of the algorithms on the 13 high-dimensional problems, where the x axis denotes the NFE (10^4) and the y axis denotes the mean best error. From the curves we can see that on all the 12 functions except

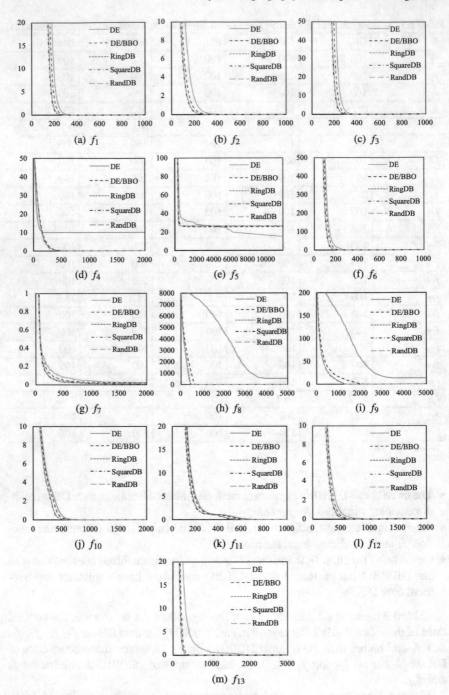

Fig. 5.1 Convergence curves of DE, DE/BBO, and Local-DE/BBO algorithms on the benchmark problems [21]

f_5, the four DE/BBO algorithms converge much faster than DE. On f_4 and f_5, the convergence speeds of the four DE/BBO versions are similar, but on the remaining 11 functions the three Local-DE/BBO algorithms converge much faster than the original DE/BBO.

In general, there is no statistically significant differences among the three Local-DE/BBO algorithms. The performance of the RingDB is slightly better than SquareDB and RandDB on some high-dimensional functions, because the contribution of the relatively large neighborhood size of SquareDB or RandDB is not so significant in hybrid DE/BBO, as DE mutation has already provided good global search ability. Nevertheless, the local topologies greatly improve the exploitation ability of BBO algorithms, among which RingDB exhibits the best performance.

5.2.3 Self-adaptive DE/BBO

Self-adaptive differential evolution (SaDE) is an improved variant of DE proposed by Qin and Suganthan [8], and its main idea is to use multiple mutation schemes, adaptively choose an appropriate mutation scheme among them and adjust the associated parameters via online learning from the preceding behavior of the already applied schemes.

Introducing this idea to DE/BBO, we propose a hybrid self-adaptive DE and BBO (HSDB) algorithm [22]. HSDB not only can choose different DE schemes dynamically, but also can choose the basic BBO migration or the improved EBO migration in a self-adaptive manner, which can greatly improve the algorithm's adaptive ability and thus balance the global exploration and local exploitation much better. We took this algorithm to participate the single-objective optimization competition in International Conference on Swarm Intelligence (ICSI) [9] in 2014 and won the first prize [13].

The HSDB method differentiates from the DE/BBO in that it does not combine DE mutation and BBO migration into an integrated operator. Instead, HSDB prefers to make more use of DE mutation for exploration in early stages and is more likely to adopt BBO migration for exploitation in later stages. The key to realize this is a control parameter named maturity index, denoted by v, which increases with the generation: The higher (lower) the v, the more likely the algorithm is to conduct migration (mutation). A simple approach is to linearly increase v from an initial value v_{min} to a final value v_{max} as:

$$v = v_{min} + \frac{t}{t_{max}}(v_{max} - v_{min}) \tag{5.1}$$

where t is the current generation number (or NFE) and t_{max} is the maximum generation number (or MNFE) of the algorithm.

At each generation, each habitat has a probability of v of being modified by BBO migration and a probability of $(1 - v)$ of being modified DE mutation. Moreover, we embed a set of DE mutation schemes (denoted by S_{DE}) and a set of BBO migration schemes (denoted by S_{BBO}) in HSDB, for each scheme record its success and failure numbers within a fixed number L_p of previous generations, and calculate its success rate in a similar way as SaDE [8]. However, in HSDB, we calculate the success rates for DE mutation schemes and BBO migration schemes independently. At beginning, every scheme is assigned with an equal selection probability; at each generation $G \geq L_p$, the probability of choosing the kth DE mutation scheme and that of the kth BBO migration scheme is, respectively, updated as follows:

$$p_k(G) = s_k(G)/\sum\nolimits_{k \in S_{DE}} s_k(G) \tag{5.2}$$

$$p_{k'}(G) = s_{k'}(G)/\sum\nolimits_{k' \in S_{BBO}} s_{k'}(G) \tag{5.3}$$

In [22], HSDB embeds two DE mutation schemes including DE/rand /1/bin and DE/rand-to-best/2/bin, and two BBO migration schemes including the original clonal migration and the local migration [4, 20]. This is because the two DE mutation schemes exhibit good exploration ability and the two BBO migration schemes have more exploitation ability than others. These operation schemes can be seen in Eqs. (1.24), (1.29), and (4.1).

To support local migration, HSDB employs the local random topology as in Sect. 3.3 and uses the procedure show in Algorithm 5.4 to set the neighborhood structure for the population. The procedure is similar with Algorithm 3.3, but in HSDB each habitat has at least one neighbor; that is, there is no isolated habitat. If there is no new better solution found after N_I (which is usually set between 1 and 6) iterations, the algorithm will reset the neighborhood structure.

Algorithm 5.4: The procedure of setting the neighborhood topology in HSDB algorithm

---

1 Initialize an $N * N$ matrix **Link**;
2 Let $p = K_N/(N-1)$;
3 **for** $i = 1$ *to* N **do**
4   let $iso = true$;
5   **for** $j = 1$ *to* N **do**
6   **if** rand() $< p$ **then** set **Link**$(i, j) = $ **Link**$(j, i) = 1, iso = false$;
7   **else** set **Link**$(i, j) = $ **Link**$(j, i) = 0$;
8   **if** iso **then**
9   Let $j = rand(1, N)$;
10   **while** $j = i$ **do** $j = $ rand$(1, N)$;
11   set **Link**$(i, j) = $ **Link**$(j, i) = 1$;

The local migration is performed as described in Algorithm 3.4. For DE crossover, the rate c_r is replaced by the immigration rate λ_i, and thus, the DE crossover operation in Eq. (1.25) is replaced by Eq. (5.4):

$$u_{ij} = \begin{cases} v_{ij}, & (\text{rand}() \leq \lambda_i) \vee (j = r_i) \\ x_{ij}, & \text{otherwise} \end{cases} \tag{5.4}$$

For consistency, the parameter F in Eqs. (1.24) and (1.29) and the parameter α in Eq. (4.1) are denoted by variable F, the value of which is produced by a Gaussian distribution with mean F_μ and variance F_σ as

$$F = \text{norm}(F_\mu, F_\sigma) \tag{5.5}$$

In HSDB, we set F_μ to 0.5 and F_σ to 0.3, and limit the value F in the range [0.05, 0.95] as

$$F = \begin{cases} 0.05, & F < 0.05 \\ 0.95, & F > 0.95 \\ F, & 0.05 \leq F \leq 0.95 \end{cases} \tag{5.6}$$

Algorithm 5.5 presents the algorithmic framework of HSDB.

Algorithm 5.5: The HSDB algorithm

1 Initialize a population of solutions;
2 Evaluate the fitness of the solutions;
3 Initialize the neighborhood structure by the Algorithm 5.4;
4 Initialize the set of DE mutation and BBO migration schemes, and assign each scheme with same selection probability;
5 **while** *the stop criterion is not satisfied* **do**
6   Calculate the migration rates of the solutions;
7   Update the maturity index ν according to Eq. (5.1);
8   **forall the** *solution* $\mathbf{x}_i$ *in the population* **do**
9   **if** rand() $\leq \nu$ **then** select a DE mutation operator k with probability $\propto p_k(G)$;
10   **else** Select a BBO migration operator k' with probability $\propto p_{k'}(G)$ Generate a mutation vector $\mathbf{v}_i$;
11   Generate a trial vector $\mathbf{u}_i$;
12   Evaluate the fitness of $\mathbf{u}_i$, and keep the better between $\mathbf{u}_i$ and $\mathbf{x}_i$ in the population;
13   **if** $G \geq L_p$ **then** Update all $p_k(G)$ and $p_{k'}(G)$ according to Eqs. (5.2) and (5.3);
14   **if** *there is no new best solution found after N_I iterations* **then**
15   reset the neighborhood structure by Algorithm 5.4;

16 **return** *the optimal solution found so far;*

We test the HSDB algorithm on the ICSI 2014 benchmark set [12], which consists of 30 high-dimensional problems, denoted as $f_1 \sim f_{30}$, all shifted and rotated. Since all the test problems are considered as black box problems, we do not tune parameters for each problem; instead, we use a general parameter setting as $\nu_{\min} = 0.05$, $\nu_{\max} = 0.95$, $L_p = 30$, $N_I = 3$, $K_N = 3$, and $N = 50$.

As requested by the ICSI 2014 competition session, the algorithm has been tested on each problem with dimensions of 2, 10, and 30, respectively, the search space is

Table 5.4 Experimental results of HSDB on 2-D problems [22]

f	Maximum	Minimum	Mean	Median	Std
f_1	0.00E+00	0.00E+00	0.00E+00	0.00E+00	0.00E+00
f_2	2.98E–27	0.00E+00	1.13E–28	0.00E+00	4.18E–28
f_3	1.67E+00	1.67E+00	1.67E+00	1.67E+00	1.12E+02
f_4	0.00E+00	0.00E+00	0.00E+00	0.00E+00	0.00E+00
f_5	1.52E+02	8.00E–20	2.45E–18	1.53E–18	2.44E–18
f_6	1.97E–31	1.97E–31	1.97E–31	1.97E–31	0.00E+00
f_7	1.21E+02	0.00E+00	6.44E–18	9.65E–18	5.26E–18
f_8	8.88E+02	8.88E+02	8.88E+02	8.88E+02	9.96E–32
f_9	–4.00E+00	–4.00E+00	–4.00E+00	–4.00E+00	2.2.4E–15
f_{10}	–9.80E–01	–1.00E+00	–9.96E–01	–1.00E+00	5.86E+02
f_{11}	0.00E+00	0.00E+00	0.00E+00	0.00E+00	0.00E+00
f_{12}	5.20E+01	4.86E+01	4.90E+01	4.86E+01	8.74E–01
f_{13}	1.94E–02	0.00E+00	7.00E+00	0.00E+00	9.15E–03
f_{14}	6.99E–02	1.00E–11	5.90E–03	9.41E–04	1.06E–02
f_{15}	3.60E–03	3.91E–08	7.92E–04	3.59E–04	9.40E–04
f_{16}	0.00E+00	0.00E+00	0.00E+00	0.00E+00	0.00E+00
f_{17}	0.00E+00	0.00E+00	0.00E+00	0.00E+00	0.00E+00
f_{18}	–8.38E+02	–8.38E+02	–8.38E+02	–8.38E+02	4.59E–13
f_{19}	1.35E–32	1.35E–32	1.35E–32	1.35E–32	1.11E–47
f_{20}	5.77E–23	0.00E+00	0.00E+00	0.00E+00	0.00E+00
f_{21}	0.00E+00	0.00E+00	0.00E+00	0.00E+00	0.00E+00
f_{22}	0.00E+00	0.00E+00	0.00E+00	0.00E+00	0.00E+00
f_{23}	8.97E+00	8.97E+00	8.97E+00	8.97E+00	7.64E–15
f_{24}	–4.53E+00	–4.53E+00	–4.53E+00	–4.53E+00	6.28E–15
f_{25}	–1.39E+00	–1.71E+00	–1.60E+00	–1.71E+00	1.54E–01
f_{26}	6.67E–01	5.71E–01	5.87E–01	5.71E–01	2.79E–02
f_{27}	–3.74E+07	–3.74E+07	–3.74E+07	–3.74E+07	4.01E+02
f_{28}	–4.63E+00	–5.83E+00	–5.22E+00	–5.25E+00	2.63E–01
f_{29}	2.00E+01	2.00E+01	2.00E+01	2.00E+01	8.92E–03
f_{30}	1.01E+00	2.67E–01	4.00E–01	2.70E–01	2.71E–01

$[100, 100]^D$, the MNFE is set as 10000D (where D denotes the problem dimension), and 51 simulation runs have been performed on each problem instance. Any function value smaller than 2^{-52} is considered as zero.

In our experimental environment, the mean time for executing the benchmark program of ICSI 2014 competition over 5 runs is $T1 = 48.62$ s, and the mean time of HSDB on function f_9 and $D = 30$ over 5 runs is $T2 = 116.02$ s, and the time complexity is evaluated by the ratio $(T2 - T1)/T1 = 1.386$.

Tables 5.4, 5.5, and 5.6 present the computational results of HSDB on 2-D, 10-D, and 30-D problems, respectively. As we can see, on the simple 2-D prob-

Table 5.5 Experimental results of HSDB on 10-D problems [22]

f	Maximum	Minimum	Mean	Median	Std
f_1	1.35E+02	5.03E–02	8.23E+00	2.28E+00	1.87E+01
f_2	2.64E+02	2.03E–01	3.26E+01	5.43E+00	5.33E+01
f_3	1.70E+02	1.70E+02	1.70E+02	1.70E+02	1.77E–02
f_4	2.96E+00	6.54E–04	3.45E–01	1.55E–01	4.67E–01
f_5	1.27E–01	1.38E–05	1.14E–02	8.59E–04	2.33E–02
f_6	9.57E+00	4.97E–01	5.02E+00	5.95E+00	2.40E+00
f_7	1.15E–03	1.00E–12	1.46E–04	1.96E–05	2.74E–04
f_8	1.16E+00	5.66E–13	1.71E–01	5.16E–03	3.71E–01
f_9	–1.95E+01	–2.00E+01	–1.98E+01	–1.99E+01	1.35E–01
f_{10}	–7.42E+00	–8.76E+00	–8.34E+00	–8.37E+00	2.85E–01
f_{11}	4.56E+00	0.00E+00	1.09E+00	1.51E–02	1.82E+00
f_{12}	5.40E–01	3.76E–01	4.52E–01	4.47E–01	4.17E–02
f_{13}	6.35E–01	1.55E–01	3.99E–01	4.02E–01	1.37E–01
f_{14}	9.91E–02	1.01E–02	5.17E–02	5.25E–02	2.19E–02
f_{15}	9.83E–02	1.74E–02	4.89E–02	4.84E–02	1.76E–02
f_{16}	4.01E–02	1.14E–08	6.05E–03	1.20E–03	9.83E–03
f_{17}	4.72E–03	3.06E–06	8.07E–04	2.72E–04	1.03E–03
f_{18}	–2.02E+03	–4.14E+03	–2.30E+03	–2.02E+03	6.89E+02
f_{19}	1.62E–03	5.20E–21	6.90E–04	9.55E–04	6.29E–04
f_{20}	3.68E–02	4.98E–06	5.61E–03	1.90E–03	8.75E–03
f_{21}	9.99E–02	9.99E–02	9.99E–02	9.99E–02	7.70E–11
f_{22}	1.08E–02	5.32E–31	8.48E–04	3.56E–05	1.90E–03
f_{23}	2.37E+01	1.55E+01	1.94E+01	1.90E+01	2.01E+00
f_{24}	–3.47E+01	–3.56E+01	–3.53E+01	–3.52E+01	2.40E–01
f_{25}	4.30E+01	4.29E+01	4.29E+01	4.29E+01	2.80E–03
f_{26}	7.59E–03	5.90E–03	6.72E–03	6.70E–03	4.87E–04
f_{27}	–6.34E+07	–1.60E+08	–1.02E+08	–1.07E+08	1.70E+07
f_{28}	–5.04E+00	–5.85E+00	–5.38E+00	–5.40E+00	1.72E–01
f_{29}	2.00E+01	2.00E+01	2.00E+01	2.00E+01	9.31E–04
f_{30}	1.01E+00	1.01E+00	1.01E+00	1.01E+00	1.86E–10

lems, HSDB achieves a function error value less than 2^{-52} on 21 problems. With the increase of dimension, the performance of HSDB decreases on most problems. This phenomenon, known as the curse of dimensionality, is serious on several problems such as f_1 and f_2. However, an interesting finding is that the solution accuracy of the algorithm increases on some special problems including f_{12} and f_{26}.

We compare the performance of HSDB with the DE algorithm using DE/rand/1/ bin, SaDE, and B-BBO on the 30-dimensional problem set, and conduct nonparametric Wilcoxon rank-sum tests on the results of HSDB and that of the other three

Table 5.6 Experimental results of HSDB on 30-D problems [22]

f	Maximum	Minimum	Mean	Median	Std
f_1	1.69E+05	4.08E+03	7.67E+04	7.80E+04	4.15E+04
f_2	1.60E+04	1.34E+03	5.18E+03	3.71E+03	3.27E+03
f_3	4.53E+03	4.51E+03	4.52E+03	4.52E+03	5.11E+00
f_4	1.33E+01	1.59E+00	7.45E+00	7.31E+00	2.71E+00
f_5	7.87E-02	8.39E-03	3.77E-02	3.66E-02	1.72E-02
f_6	3.03E+01	2.82E+01	2.91E+01	2.90E+01	5.15E-01
f_7	1.01E-02	3.23E-03	7.37E-03	7.84E-03	1.72E-03
f_8	1.60E+00	1.86E-01	9.06E-01	9.91E-01	4.42E-01
f_9	−5.76E+01	−5.91E+01	−5.83E+01	−5.82E+01	3.35E-01
f_{10}	−1.99E+01	−2.79E+01	−2.48E+01	−2.54E+01	1.80E+00
f_{11}	8.90E-02	4.55E-04	2.98E-02	2.85E-02	1.82E-02
f_{12}	1.47E-02	1.22E-02	1.43E-02	1.44E-02	4.42E-04
f_{13}	4.02E+00	1.40E+00	3.03E+00	3.14E+00	6.88E-01
f_{14}	1.03E-01	2.53E-02	7.29E-02	7.43E-02	1.96E-02
f_{15}	1.70E-01	2.04E-02	9.48E-02	9.36E-02	3.40E-02
f_{16}	3.33E-01	8.08E-02	2.21E-01	2.24E-01	5.23E-02
f_{17}	3.93E+00	4.31E-01	2.02E+00	2.10E+00	8.17E-01
f_{18}	−3.69E+03	−6.06E+03	−5.79E+03	−6.06E+03	7.11E+02
f_{19}	8.67E-03	2.00E-03	5.05E-03	4.81E-03	1.77E-03
f_{20}	7.45E-01	9.08E-06	2.07E-01	1.11E-01	2.27E-01
f_{21}	3.00E-01	9.99E-02	2.13E-01	2.00E-01	5.30E-02
f_{22}	2.15E-01	1.01E-02	1.01E-01	9.68E-02	5.44E-02
f_{23}	5.68E+01	3.09E+01	4.43E+01	4.32E+01	7.31E+00
f_{24}	−1.13E+02	−1.16E+02	−1.15E+02	−1.15E+02	7.39E-01
f_{25}	1.13E+03	1.13E+03	1.13E+03	1.13E+03	2.06E+00
f_{26}	4.48E-04	4.12E-04	4.33E-04	4.35E-04	8.73E-06
f_{27}	−2.99E+08	−5.25E+08	−4.76E+08	−5.21E+08	8.72E+07
f_{28}	−5.24E+00	−5.81E+00	−5.84E+00	−5.84E+00	1.12E-01
f_{29}	2.00E+01	2.00E+01	2.00E+01	2.00E+01	1.53E-04
f_{30}	1.04E+00	1.01E+00	1.02E+00	1.01E+00	1.34E-02

methods. Table 5.7 presents the comparison results, where a superscript $^+$ denotes HSDB has statistically improvement over the compared algorithms and $^-$ vice verse (at 95% confidence). The results show that:

- HSDB has obvious performance advantage over DE on 28 functions, and DE has performance advantage over HSDB only on function f_{26}. There is no statistical differences between them on function f_{28}.

Table 5.7 Comparison of HSDB with three other algorithms on 30-D problems [22]

f	DE		SaDE		B-BBO	
	Median	Std	Median	Std	Median	Std
f_1	1.73E+06$^+$	1.10E+06	1.98E+04$^-$	2.95E+04	3.98E+06$^+$	9.84E+05
f_2	8.28E+05$^+$	1.87E+05	2.58E+04$^+$	8.72E+03	5.72E+04$^+$	2.18E+04
f_3	5.26E+03$^-$	7.97E+02	4.51E+03	5.27E+00	4.84E+03$^+$	1.30E+02
f_4	3.72E+02$^+$	6.08E+01	1.76E+02$^+$	3.06E+01	1.36E+02$^+$	3.23E+01
f_5	1.39E+00$^+$	9.06E–01	2.31E–02$^-$	1.36E–02	1.64E+00$^+$	3.71E–01
f_6	3.77E+01$^+$	9.47E+00	2.84E+01$^-$	4.02E–01	6.91E+01$^+$	1.79E+01
f_7	3.98E–02$^+$	3.87E–03	8.74E–03$^+$	3.81E–03	3.10E–02$^+$	1.38E–02
f_8	2.29E+00$^+$	4.90E–01	9.81E–01	6.12E–01	3.37E+00$^+$	4.41E–01
f_9	–5.61E+01$^+$	7.52E–01	–5.84E+01	3.60E–01	–5.58E+01$^+$	5.06E–01
f_{10}	–1.03E+01$^+$	1.12E+00	–1.84E+01$^+$	1.10E+00	–2.42E+01$^+$	2.43E+00
f_{11}	2.82E–01$^+$	2.92E–01	1.21E–02$^-$	1.14E–02	2.73E+00$^+$	8.30E–01
f_{12}	1.53E–02$^+$	1.57E–04	1.48E–02$^+$	1.87E–04	1.46E–02$^+$	6.13E–04
f_{13}	7.40E+00$^+$	4.03E–01	5.16E+00$^+$	5.64E–01	4.79E+00$^+$	5.59E–01
f_{14}	1.40E–01$^+$	2.29E–02	9.16E–02$^+$	1.97E–02	8.37E–02$^+$	2.02E–02
f_{15}	1.46E–01$^+$	4.37E–02	1.19E–01$^+$	3.67E–02	4.78E–01$^+$	8.24E–02
f_{16}	5.63E–01$^+$	1.50E–01	2.07E–01	8.51E–02	1.01E+00$^+$	2.24E–01
f_{17}	1.69E+02$^+$	2.91E+01	1.19E+01$^+$	2.83E+00	1.14E+01$^+$	3.15E+00
f_{18}	–6.05E+03$^+$	1.66E+00	–6.06E+03$^+$	7.60E+02	–3.66E+03$^+$	1.91E+03
f_{19}	2.85E–01$^+$	1.39E–01	2.35E–02$^+$	1.53E–02	1.90E–02$^+$	8.13E–03
f_{20}	6.04E+02$^+$	2.07E+02	7.82E+01$^+$	3.06E+01	1.89E+01$^+$	7.50E+00
f_{21}	7.78E–01$^+$	2.18E–01	2.00E–01	6.16E–02	8.00E–01$^+$	2.79E–01
f_{22}	8.45E–01$^+$	6.23E–01	1.05E–02$^-$	1.94E–02	7.75E+00$^+$	2.32E+00
f_{23}	9.28E+01$^+$	5.00E+00	6.98E+01$^+$	4.62E+00	6.69E+01$^+$	4.92E+00
f_{24}	–1.10E+02$^+$	1.49E+00	–1.14E+02	8.46E–01	–1.10E+02$^+$	1.24E+00
f_{25}	1.32E+03$^+$	7.89E+01	1.13E+03$^-$	1.69E+00	1.21E+03$^+$	2.57E+01
f_{26}	3.98E–04$^|$	1.19E–06	4.26E–04	1.58E–05	6.26E–04$^+$	5.67E–05
f_{27}	–4.64E+08$^+$	3.13E+07	–5.19E+08	4.89E+07	–2.69E+08$^+$	5.82E+07
f_{28}	–5.49E+00	1.26E–01	–5.51E+00	1.44E–01	–5.49E+00	1.14E–01
f_{29}	2.00E+01$^+$	1.26E–04	2.00E+01$^+$	1.18E–04	2.00E+01$^+$	1.57E–04
f_{30}	1.04E+00$^+$	1.14E–02	1.01E+00$^+$	1.34E–02	1.04E+00$^+$	8.03E–03

- HSDB has obvious performance advantage over SaDE on 15 functions, and SaDE has performance advantage over HSDB on 6 functions. There is no statistical differences between them on the rest 9 functions.
- HSDB has obvious performance advantage over B-BBO on 29 functions, and there is no statistical differences between them only on function f_{28}.

In summary, the overall performance of HSDB is better than the other three algorithms on the benchmark set.

5.3   Hybridization with Harmony Search

5.3.1   Biogeographic Harmony Search

In the original HS [2], the harmony memory consideration operator (Line 8 in Algorithm 1.11) initializes a new harmony by randomly choosing existing solution components from the HM, which limits the component diversity of the new harmony. Therefore, we propose a hybrid HS and BBO method, named biogeographic harmony search (BHS), which integrates the blended migration operator of BBO into harmony generation to promote the information sharing in the HM and enhance the solution diversity [23].

Suppose the solutions in the HM are sorted in decreasing order of fitness, we use a simple linear migration model to calculate the emigration rate μ_i and immigration rate λ_i for each ith solution as follows (where $|HM|$ denotes the memory size):

$$\lambda_i = i/|HM| \tag{5.7}$$

$$\mu_i = 1 - i/|HM| \tag{5.8}$$

When initializing a new harmony $\mathbf{x}$, at each dimension d, we randomly select a solution x_i from the HM and make the new component $x(d)$ has a probability of $(1 - \lambda_i)$ of being $x_i(d)$, and a probability of λ_i of being immigrated from another solution x_j selected according to the emigration rate. The blended migration is used in the BHS to improve component diversity:

$$x(d) = x_i(d) + \text{rand}() \cdot (x_j(d) - x_i(d)) \tag{5.9}$$

This makes the new solution component be a new value between x_i and x_j. Under the migration schema, if x_i has a high fitness value, it will be less likely to degrade since it has a low immigration rate; otherwise, it has a chance of accepting many new features from other good solutions.

Moreover, if a component has been immigrated, no pitch adjustment is further needed. In other words, pitch adjustment is only performed (with a probability of *PAR*) on a component directly taking from the HM, which can improve the local search ability of the method.

Algorithm 5.6 presents the framework of the BHS method. As can be seen, harmony memory consideration and pitch adjustment can be regarded as a complementation to migration: A low-quality solution x_i has a high immigration rate λ_i and, thus, tends to immigrate many features from memory solutions; on the contrary, a high-quality solution x_i has a low λ_i and, thus, is more likely to perform local search around itself. In consequence, the combination of harmony search and migration is expected to provide a much better balance between global exploration and local exploitation.

Algorithm 5.6: The biogeographic harmony search algorithm

---

1 Initialize a population HM of solutions;
2 **while** *the stop criterion is not satisfied* **do**
3   Sort the harmonies in decreasing order of fitness;
4   Calculate their migration rates according to Eqs. (5.7) and (5.7) ;
5   Create a new blank harmony $\mathbf{x}$;
6   **forall the** *dimension d* **do**
7   **if** rand$()$ > *HMCR* **then**
8   Set $x(d)$ as a random value in the range;
9   **else**
10   Pick a random harmony $\mathbf{x}_i$ from HM;
11   **if** rand$()$ ≤ λ_i **then**
12   Pick another harmony $\mathbf{x}_j$ from the harmony memory according to μ_j;
13   Calculate the component value according to Eq. (5.9);
14   **else**
15   Set $x(d) = x_i(d)$;
16   **if** rand$()$ ≤ *PAR* **then** adjust pitch according to Eq. (1.30);
17   Calculate the fitness of $\mathbf{x}$;
18   **if** $\mathbf{x}$ *is better than the worst harmony* **then** replace the worst harmony with x;
19 **return** *the best solution found so far;*

5.3.2 Computational Experiments

We test the BHS algorithm on a set of well-known benchmark functions, including 13 high-dimensional unimodal and multimodal problems [16] and 7 shifted and rotated problems [5]. We compare BHS with other six algorithms: the original HS [2], IHS [7], DHS [1], BBO [10], B-BBO [6], and another hybrid algorithm HSBBO [15]. We set the HM size to 10 for the three HS algorithms and BHS and the population size to 50 for the BBO and B-BBO. For the original IIS, we set $HMCR = 0.9$, $PAR = 0.3$, and $bw = 0.01$. For IHS, we set $HMCR = 0.95$, $PAR_{\max} = 0.99$, $PAR_{\min} = 0.35$, $bw_{\max} = 0.1$, and $bw_{\min} = 0.00001$. For the IISBBO, we set $HMCR = 0.9$, $PAR = 0.1$, and $bw = 0.01$. The parameters of BBO and B-BBO are set as suggested in the literature. For BHS, we use the same parameter settings and adaptive mechanisms as the IHS (note that BHS does not add any new parameter to IHS).

For a fair comparison, the maximum number of function evaluations is set to 200,000 for all algorithms. On each problem, a Monte Carlo experiment is performed 60 times, all with different random seeds. Among the 60 simulation runs, the average (mean and standard deviation), best, and worst function values at the end of each run are calculated as the experimental results.

Tables 5.8 and 5.9 present the experimental results, and the best mean function values among the seven algorithms are in boldface. In columns 4–9, a superscript † indicates that BHS has a statistically significant performance improvement over the corresponding algorithm (at a confidence level of 95%) according to the paired

Table 5.8 Results of BHS and comparative algorithms on functions f_1-f_{13}

f	Name	Metric	BBO	B-BBO	HS	IHS	DHS	HSBBO	BHS
f_1	Sphere	Mean	6.15E-01[†]	1.80E+01[†]	1.44E-04[†]	1.07E-09[†]	4.08E-03[†]	5.33E-06[†]	**2.78E-10**
		Std	2.34E-01	3.09E+00	3.93E-05	1.16E-10	1.29E-02	7.20E-06	4.18E-11
		Best	2.46E-01	1.16E+01	9.03E-05	8.27E-10	7.99E-13	7.12E-07	2.02E-10
		Worst	1.15E+00	2.54E+01	3.10E-04	1.30E-09	7.70E-02	1.63E-05	3.78E-10
f_2	Schwefel 2.22	Mean	2.14E-01[†]	5.10E-01[†]	3.22E-02[†]	1.37E-04[†]	6.19E-04[†]	2.09E-03[†]	**6.63E-05**
		Std	3.82E-02	7.07E-02	3.39E-03	8.14E-06	2.05E-03	9.08E-04	5.25E-05
		Best	1.22E-01	3.78E-01	2.44E-02	1.17E-04	2.50E-10	9.75E-04	4.77E-05
		Worst	2.86E-01	6.83E-01	4.04E-02	1.55E-04	1.08E-02	4.38E-03	7.58E-05
f_3	Schwefel 1.2	Mean	1.24E+01[†]	2.89E+02[†]	3.67E-03[†]	1.50E-08[†]	2.19E-02[†]	7.37E-05[†]	**3.91E-09**
		Std	5.78E+00	5.80E+01	1.97E-03	1.82E-09	6.67E-02	4.20E-05	6.09E-10
		Best	4.02E+00	1.69E+02	1.45E-03	9.56E-09	2.33E-11	2.99E-05	2.59E-09
		Worst	3.43E+01	4.16E+02	1.03E-02	2.00E-08	4.60E-08	9.78E-05	5.46E-09
f_4	Schwefel 2.21	Mean	2.08E+00[†]	1.03E+00[†]	2.28E+00[†]	1.25E+00[†]	2.80E+00[†]	2.37E+00[†]	**6.53E-01**
		Std	3.84E-01	1.33E-01	4.16E-01	2.25E-01	9.05E-01	2.35E-01	1.28E-01
		Best	1.19E+00	8.20E-01	1.59E+00	7.71E-01	1.25E+00	2.08E+00	3.95E-01
		Worst	2.92E+00	1.40E+00	3.29E+00	1.67E+00	5.18E+00	2.89E+00	9.56E-01
f_5	Rosenbrock	Mean	1.74E+02[†]	1.94E+03[†]	8.58E+01[†]	6.10E+01	8.04E+01[†]	8.45E+01[†]	**5.70E+01**
		Std	7.13E+01	5.75E+02	5.86E+01	3.74E+01	4.19E+01	3.60E+01	3.53E+01
		Best	9.74E+01	7.53E+02	3.58E+01	3.35E+00	8.11E+00	3.75E+01	4.25E+01
		Worst	6.31E+02	3.12E+03	3.50E+02	1.40E+02	2.01E+02	1.67E+02	1.34E+02
f_6	Step	Mean	2.17E-01[†]	1.89E+01[†]	1.83E-01[†]	**0.00+00**	5.33E-01[†]	4.96E-01[†]	**0.00E+00**
		Std	4.15E-01	3.51E+00	4.31E-01	0.00E+00	7.69E-01	5.25E-01	0.00E+00
		Best	0.00E+00	1.40E+01	0.00E+00	0.00E+00	0.00E+00	0.00E+00	0.00E+00
		Worst	1.00E+00	2.80E+01	2.00E+00	0.00E+00	3.00E+00	1.00E+00	0.00E+00
f_7	Quartic	Mean	1.42E-02[†]	1.34E-02	1.91E-02[†]	1.73E-02[†]	2.12E-02[†]	1.50E-02	**1.04E-02**
		Std	9.32E-03	8.34E-03	1.27E-02	1.25E-02	1.17E-02	1.00E-02	8.22E-03

(continued)

Table 5.8 (continued)

f	Name	Metric	BBO	B-BBO	HS	IHS	DHS	HSBBO	BHS
f_8	Generalized Schwefel 2.26	Best	1.00E-03	1.44E-03	3.55E-04	8.41E-04	2.99E-03	3.08E-03	2.01E-04
		Worst	3.56E-02	3.38E-02	4.75E-02	5.00E-02	5.08E-02	2.28E-02	2.88E-02
		Mean	9.57E-01†	1.34E+02†	9.11E-01†	1.29E-10†	5.41E+01	6.38E-05†	**3.30E-11**
		Std	4.72E-01	3.63E-01	7.44E-01	1.77E-11	3.31E+02	2.12E-04	4.75E-12
		Best	4.05E-01	7.47E-01	1.98E-01	7.95E-11	1.48E-08	3.39E-07	2.19E-11
		Worst	2.51E+00	2.19E+02	3.14E+00	1.77E-10	2.52E+03	8.13E-05	4.57E-11
f_9	Rastrigin	Mean	1.82E-01†	1.02E-01†	1.98E-02†	2.06E-07†	1.31E+02†	5.50E-03†	**5.54E-08**
		Std	6.27E-02	5.32E-02	4.04E-03	2.69E-08	1.63E+01	1.96E-03	7.76E-09
		Best	6.20E-02	3.43E-02	1.03E-02	1.28E-07	8.14E+01	9.70E-08	3.84E-08
		Worst	3.91E-01	2.71E-01	3.17E-02	2.59E-07	1.68E+02	7.22E-03	7.41E-08
f_{10}	Ackley	Mean	1.68E-01†	3.12E-02†	7.75E-03†	2.32E-05†	6.01E-03†	3.29E-05	**1.23E-05**
		Std	4.37E-02	5.64E-03	7.88E-04	1.73E-06	1.19E-02	3.76E-05	8.26E-07
		Best	8.90E-02	2.07E-02	6.38E-03	1.84E-05	1.48E-07	1.29E-05	1.02E-05
		Worst	3.38E-01	4.56E-02	9.84E-03	2.69E-05	6.32E-02	4.95E-05	1.42E-05
f_{11}	Griewank	Mean	5.75E-01†	4.49E-02†	3.30E-01†	2.59E-02	**1.06E-02**	2.38E-02†	1.97E-02
		Std	1.35E-01	2.69E-02	1.28E-01	2.85E-02	2.28E-02	1.11E-02	1.81E-02
		Best	2.79E-01	1.34E-02	9.21E-02	4.84E-11	7.14E-11	2.05E-02	9.34E-12
		Worst	9.47E-01	1.80E-01	6.65E-01	1.15E-01	1.56E-01	1.32E-01	7.01E-02
f_{12}	Generalized penalized 1	Mean	4.92E-02†	1.32E-03†	4.73E-04†	4.90E-11†	1.07E-04†	3.01E-10†	**1.13E-11**
		Std	2.75E-02	3.87E-04	1.69E-03	7.08E-12	3.38E-04	1.88E-09	1.50E-12
		Best	1.78E-02	5.90E-04	3.64E-06	2.99E-11	3.20E-12	1.57E-09	6.72E-12
		Worst	1.44E-01	2.67E-03	9.12E-03	6.38E-11	2.26E-03	6.03E-11	1.44E-11
f_{13}	Generalized penalized 2	Mean	2.21E-01†	3.07E-02†	4.42E-02†	6.04E-03†	1.59E-03	1.36E-03	**1.05E-03**
		Std	1.09E-01	1.92E-02	7.01E-02	8.58E-03	4.16E-03	3.86E-03	2.95E-03
		Best	5.74E-02	5.99E-03	1.72E-05	1.34E-10	1.80E-10	9.52E-11	2.38E-11
		Worst	5.29E-01	7.58E-02	4.74E-01	3.44E-02	2.20E-02	1.45E-02	1.10E-02

Table 5.9 Results of BHS and comparative algorithms on functions f_{14}–f_{20}

f	Name	Metric	BBO	B-BBO	HS	IHS	DHS	HSBBO	BHS
f_{14}	SR high-conditioned Elliptic	Mean	4.87E+06†	1.65E+07†	3.35E+06†	2.77E+06†	1.12E+07†	3.29E+06†	**2.21E+06**
		Std	2.19E+06	2.34E+06	1.64E+06	1.28E+06	6.27E+06	2.00E+06	1.01E+06
		Best	1.95E+06	1.17E+07	1.38E+06	7.24E+05	1.89E+06	1.45E+06	4.35E+05
		Worst	1.19E+07	2.25E+07	8.77E+06	8.06E+06	3.43E+07	6.39E+06	4.29E+06
f_{15}	SR Bent Cigar	Mean	2.42E+06†	1.08E+09†	2.16E+05†	9.57E+04†	2.25E+04	1.18E+05†	**1.30E+04**
		Std	1.50E+06	1.51E+08	7.15E+05	3.02E+05	9.31E+04	9.68E+04	6.57E+04
		Best	6.67E+05	7.00E+08	4.66E+02	2.02E+02	2.01E+02	1.09E+05	2.08E+02
		Worst	9.60E+06	1.42E+09	3.33E+06	1.82E+06	7.16E+05	5.60E+05	4.86E+05
f_{16}	R Discus	Mean	2.69E+04†	4.15E+04†	2.27E+04	3.19E+04†	2.77E+04†	3.19E+04†	**1.97E+04**
		Std	1.43E+04	4.49E+03	1.19E+04	1.72E+04	5.90E+03	1.55E+04	8.75E+03
		Best	5.24E+03	2.98E+04	5.28E+03	3.32E+03	1.20E+04	7.02E+03	4.45E+03
		Worst	7.83E+04	5.48E+04	6.40E+04	8.41E+04	4.66E+04	5.81E+04	4.00E+04
f_{17}	SR Rosenbrock	Mean	4.89E+02†	6.72E+02†	4.82E+02	4.84E+02	**4.70E+02**	4.80E+02	4.80E+02
		Std	2.81E+01	3.46E+01	2.70E+01	2.90E+01	1.28E+01	2.32E+02	1.17E+01
		Best	4.31E+02	6.06E+02	4.22E+02	4.34E+02	4.33E+02	4.33E+02	4.38E+02
		Worst	5.75E+02	7.37E+02	5.45E+02	5.61E+02	4.95E+02	5.36E+02	5.00E+02
f_{18}	SR Griewank	Mean	3.57E+06†	1.93E+09†	1.00E+04	7.00E+02†	7.54E+03†	9.68E+03†	**7.00E+02**
		Std	1.62E+06	2.42E+08	4.70E+04	2.88E+04	1.57E+04	1.27E+04	9.77E-05
		Best	6.89E+05	1.49E+09	9.59E+02	7.00E+02	7.00E+02	7.00E+02	7.00E+02
		Worst	8.41E+06	2.53E+09	2.70E+05	7.00E+02	8.62E+04	5.05E+04	7.00E+02
f_{19}	S Rastrigin	Mean	8.00E+02†	8.48E+02†	8.00E+02†	8.00E+02†	9.24E+02†	8.00E+02†	**8.00E+02**
		Std	7.27E-02	5.70E+00	1.43E-05	7.76E-11	1.93E+01	1.58E-10	2.03E-11
		Best	8.00E+02	8.36E+02	8.00E+02	8.00E+02	8.72E+02	8.00E+02	8.00E+02
		Worst	8.00E+02	8.63E+02	8.00E+02	8.00E+02	9.57E+02	8.00E+02	8.00E+02
f_{20}	SR Rastrigin	Mean	9.49E+02†	9.95E+02†	9.58E+02†	9.79E+02†	1.11E+03†	9.85E+02†	**9.23E+02**
		Std	1.11E+01	1.28E+01	1.27E+01	2.03E+01	1.45E+01	3.87E+01	5.30E+00
		Best	9.24E+02	9.71E+02	9.32E+02	9.35E+02	1.08E+03	9.33E+02	9.14E+02
		Worst	9.81E+02	1.02E+03	9.89E+02	1.03E+03	1.13E+03	9.90E+02	9.38E+02

t-tests. As we can see, BHS achieves the best results on 18 of the 20 functions. Only on functions f_{11} and f_{17}, the mean values of DHS are better than BHS. Even so, BHS performs better than the other five algorithms and is ranked the second place on the two functions. Statistical tests show that the performance of BHS is significantly better than BBO on all the 20 functions, better than B-BBO on 19 functions, better than HS on 17 functions, better than IHS and HSBBO on 16 functions, and better than DHS on 15 functions.

In summary, the overall performance of BHS is the best among the seven algorithms on the benchmark suite. In particular, as a combination of B-BBO and IHS, BHS never performs worse than B-BBO and IHS on any problem, which demonstrates the effectiveness of the hybridization of the two algorithms.

5.4 Hybridization with Fireworks Algorithm

5.4.1 A Hybrid BBO and FWA Algorithm

BBO excels in local exploitation, while FWA [14] is strong in global exploration and fast convergence, so they can complement each other. In [17], we propose a hybrid BBO and FWA algorithm, denoted by BBO_FWA. As described in Sect. 1.4.8, EFWA [19] is a major improvement and the current norm of FWA, and our BBO_FWA is essentially a combination of EFWA and BBO.

The key idea of BBO_FWA is based on the observation that the BBO migration operator and the EFWA explosion operator both have their advantages and disadvantages:

- EFWA explosion achieves a good balance between exploration and exploitation based on the control of explosion amplitude and spark number. However, information sharing among the solutions is poor, and an explosion generating a number of sparks is computationally expensive.
- BBO migration is computationally cheap, and it is quite effective in information sharing among the solutions and contributes greatly to extensive local search.

BBO_FWA combines their advantages by using a migration probability ρ: At each generation, each firework has a probability of ρ of being evolved by the migration operator and a probability of $(1 - \rho)$ of being evolved by explosion. In general, for a problem with a complex objective function, we prefer to set a large ρ value to reduce the number of sparks to be generated and evaluated and thus alleviate the computational burden. A large ρ can also enhance the solution diversity and thus improve the search ability for multimodal problem. In contrast, a small ρ is more suitable for those problems whose optima are located in very narrow or sharp ridges, as the explosion operator enables the exploitation along multiple directions and thus decreases the chance of skipping the optima. Empirically, we suggest setting the value of ρ in the range between 0.5 and 0.8.

Since BBO migration can enhance the information sharing in the population to enrich the solution diversity and EFWA Gaussian mutation also shares information of the current best with other solutions, BBO_FWA does not employ the elite strategy that always put the best known solution into the next generation. Instead, the best solution is recorded outside the population. Algorithm 5.7 presents the framework of BBO_FWA.

Algorithm 5.7: The hybrid BBO_FWA algorithm

---

1 Initialize a population of N fireworks;
2 **while** *the stop criterion is not satisfied* **do**
3   Evaluate the fitness of the solutions and update the best solution found;
4   Calculate the migration rates of the solutions;
5   **forall the** *solution* $\mathbf{x}_i$ *in the population* **do**
6   **if** *the* rand() $< \rho$ **then**
7   Perform the migration according to Algorithm 2.1;
8   With a mutation probability, mutate $\mathbf{x}_i$ according to (1.35);
9   **else**
10   Calculate the number s_i of sparks according to Eq. (1.32);
11   **for** $k = 1$ *to* s_i **do**
12   Use EFWA explosion [19] to generate a spark $\mathbf{x}_j$;
13   With a mutation probability, mutate $\mathbf{x}_j$ according to (1.35);
14   Randomly select N solutions from all fireworks and sparks for the next generation;
15 **return** *the best solution found so far;*

5.4.2 Computational Experiments

We select the 13 high-dimensional benchmark functions from [16] and use 30-dimensional functions for test. For f_8 whose optimum value is not zero, we add a constant to the function expression to make its optimum value zero.

In the experiments, we first test the performance of BBO_FWA with ρ ranges from 0 to 1 with an interval of 0.05. For each ρ value, we run the algorithm 60 times on each function, and record the mean best function error value. The results reveal that on most of problems, BBO_FWA obtains the minimum mean errors when $\rho \in [0.6, 0.75]$. Therefore, we retest BBO_FWA with ρ ranges from 0.6 to 0.75 with an interval of 0.01 and record the mean best values obtained over the 60 runs. Combining the results, we present the best value of ρ with which BBO_FWA obtains the minimum mean error on each problem in Table 5.10. Note that for f_6, any ρ value in the range of [0.3, 0.95] makes the algorithm obtain the global optimum.

Figure 5.2 shows the variation of the mean bests with ρ, where the y axis values are set as $\log(f/f^*)$, i.e., the natural logarithm of the ratio of the current mean best

Table 5.10 Best ρ of BBO_FWA value for each benchmark problem. ©[2014] IEEE. Reprinted, with permission, from Ref. [17]

Function	f_1	f_2	f_3	f_4	f_5	f_6	f_7	f_8	f_9	f_{10}	f_{11}	f_{12}	f_{13}
ρ	0.7	0.8	0.67	0.55	0.7	0.3 ~ 0.95	0.75	0.95	0.75	0.67	0.45	0.73	0.68

Fig. 5.2 Variation of BBO_FWA performance with the value of ρ. ©[2014] IEEE. Reprinted, with permission, from Ref. [17]

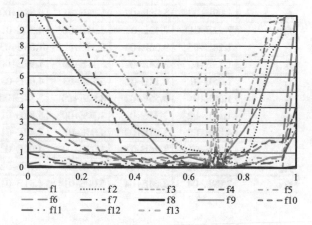

to the optimum. As we can see, the relation of the algorithm performance with the value of ρ is generally nonlinear and non-monotonic. Note that $\rho = 0$ (i.e., the only use of explosion) or $\rho = 1$ (i.e., the only use of migration) always leads to the worst results. Thus, the combination of two operators is preferred. Roughly, we suggest using a migration probability in the range of [0.6, 0.75] for most unknown problems. However, in order to achieve more precise results, fine-tuning of ρ is needed.

We then compare BBO_FWA with the basic BBO and EFWA on the benchmark problems. For the BBO, we set $N = 50$, $I = E = 1$, and for the EFWA, we set $N = 5$, $\widehat{A} = 40$, $M_e = 50$, $M_g - 5$, $s_{min} = 2$, $s_{max} = 40$, $A_{init} = 0.02(x_{max,k} - x_{min,k})$, $A_{init} = 0.001(x_{max,k} - x_{min,k})$, for the BBO_FWA, we set $N = 50$ and inherit other parameters settings of BBO and EFWA. For the sake of fairness, we use a general setting of $\rho = 0.68$ in the comparative experiments.

On each test problem, we respectively run three algorithms for 60 times with different random seeds and compute mean best values averaged over the 60 runs. Table 5.11 presents the results of the algorithms, where the bold values indicate the minimum mean bests among the three algorithms, and in columns 2 and 3, a superscript "+" indicates that BBO_FWA has significant performance improvement over EFWA or BBO and "−" vice versa (at 95% confidence level, according to paired t-tests). The results show that the general performance of BBO_FWA is much better than EFWA and BBO. In particular, BBO_FWA has statistically significant improvement over EFWA on all the problems, which shows that the integration of migration operator can effectively improve the performance of the EFWA explosion operator. Among the 13 test problems, BBO_FWA obtains the minimum mean bests on 11

Table 5.11 Experimental results of EFWA, BBO, and BBO_FWA. ©[2014] IEEE. Reprinted, with permission, from Ref. [17]

No	EFWA		BBO		BBO_FWA	
	Mean	Std	Mean	Std	Mean	Std
f_1	2.56E+01$^+$	7.56E+00	3.49E+00$^+$	1.53E+00	**1.92E–13**	5.49E–13
f_2	5.03E+00$^+$	2.74E+00	6.94E–01$^+$	1.15E–01	**3.95E–10**	2.63E–10
f_3	1.09E+03$^+$	5.08E+02	1.28E+01$^+$	6.84E+00	**1.62E–15**	1.05E–14
f_4	7.22E–01$^+$	7.42E–01	2.91E+00$^+$	4.30E–01	**3.66E–02**	7.65E–02
f_5	1.24E+05$^+$	9.06E+04	3.53E+02$^+$	1.06E+03	**5.54E+01**	8.62E+01
f_6	1.13E+03$^+$	3.24E+02	3.17E+00$^+$	1.97E+00	**0.00E+00**	0.00E+00
f_7	1.75E–01$^+$	7.14E–02	**5.72E–03**	4.07E–03	7.23E–03	4.51E–03
f_8	1.00E+04$^+$	9.55E+02	**8.92E+00**$^-$	3.34E+00	9.19E+03	9.22E+02
f_9	1.11E+02$^+$	2.31E+01	1.70E+00	5.88E–01	**1.63E+00**	9.64E–01
f_{10}	4.89E+00	1.01E+00	9.04E–01	2.42E–01	**3.09E–10**	2.78E–10
f_{11}	2.31E–02$^+$	1.26E–02	1.00E+00$^+$	4.02E–02	**1.47E–02**	1.54E–02
f_{12}	2.52E+02$^+$	5.89E+02	9.14E–02$^+$	4.74E–02	**2.51E–02**	7.79E–02
f_{13}	2.09E+04$^+$	2.64E+04	5.00E–01$^+$	1.91E–01	**9.16E–04**	3.06E–03

problems, and BBO has the minimum mean bests only on f_7 and f_8. Moreover, BBO only has significant improvement over BBO_FWA on f_7, and on f_8, there is no statistically significant difference between BBO and BBO_FWA. On the contrary, BBO_FWA has significant improvement over BBO on 10 test problems. This also shows that the integrated explosion/migration of BBO_FWA can perform more effective search than BBO migration.

Figure 5.3 presents the convergence curves of the algorithms on the 13 problems. BBO_FWA converges faster than the other two algorithms on most of the problems including f_1–f_3, f_5–f_7, f_9, and f_{11}–f_{13}. BBO and EFWA, respectively, converge faster on f_4 and f_{10} in early stages, but both are overtaken by BBO_FWA in later stages. In particular, the convergence processes of EFWA are far behind BBO_FWA on complex multimodal problems f_6, f_{12}, and f_{13} because it is easy to be trapped in local optima. In fact, on these functions EFWA can sometimes get similar results to BBO_FWA, but it also occasionally obtains very large error values, which greatly degrades the average performance over the 60 runs. The BBO_FWA's advantages in convergence speed demonstrate that the integration of migration to EFWA can greatly improve solution diversity and thus effectively avoid premature convergence. On the other hand, on many problems the convergence curves of BBO and BBO_FWA have similar shapes, but in general BBO_FWA converges faster and reaches better results. This also demonstrates that the combination of normal explosion operator can improve the exploration ability of the migration operator, resulting in much more precise optima. In summary, the hybrid explosion and migration of BBO_FWA can achieve a much better balance between solution diversification and intensification than both BBO and EFWA.

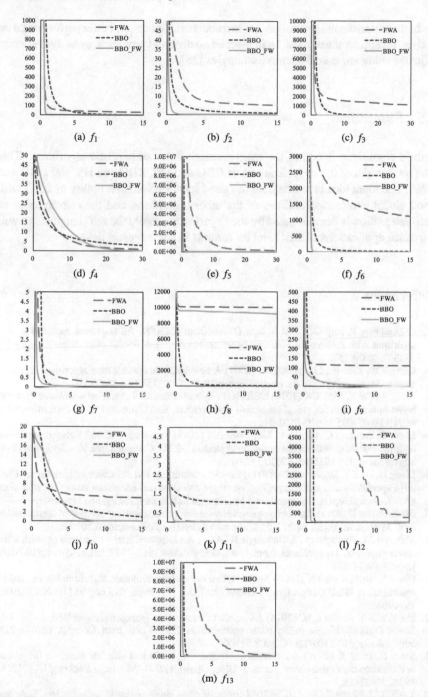

Fig. 5.3 Convergence curves of EFWA, BBO, and BBO_FWA on the benchmark problems. ©[2014] IEEE. Reprinted, with permission, from Ref. [17]

In the experiments, ρ is simply set to a constant. To achieve better performance on real-world applications, ρ can be fine-tuned on different problems, or be dynamically adjusted using some self-adaptive strategies [18].

5.5   Summary

Hybrid algorithms is one of the main research fields of heuristic algorithms. This chapter introduces the hybridizations of BBO and DE, BBO and HS, and BBO and FWA. The main idea is to integrate the good local exploitation ability of BBO with good global exploration abilities of the other algorithms, and the effectiveness of such integration is demonstrated by the experiments. From the next chapter, we will introduce applications of BBO and its variants in some typical fields.

References

1. Chakraborty P, Roy GG, Das S, Jain D, Abraham A (2009) An improved harmony search algorithm with differential mutation operator. Fundam. Inf. 95:401–426. https://doi.org/10.3233/FI-2009-157
2. Geem ZW, Kim JH, Loganathan G (2001) A new heuristic optimization algorithm: harmony search. Simulation 76:60–68. https://doi.org/10.1177/003754970107600201
3. Gong W, Cai Z, Ling CX (2010) DE/BBO: a hybrid differential evolution with biogeography-based optimization for global numerical optimization. Soft Comput. 15:645–665. https://doi.org/10.1007/s00500-010-0591-1
4. Ilhem B, Amitava C, Patrick S, Mohamed AN (2011) Two-stage update biogeography-based optimization using differential evolution algorithm DBBO. Comput. Oper. Res. 38:1188–1198. https://doi.org/10.1016/j.cor.2010.11.004
5. Liang JJ, Qu BY, Suganthan PN (2014) Problem definitions and evaluation criteria for the CEC 2014 special session and competition on single objective real-parameter numerical optimization. Technical report, Computational Intelligence Laboratory, Zhengzhou University
6. Ma H, Simon D (2011) Blended biogeography-based optimization for constrained optimization. Eng. Appl. Artif. Intell. 24:517–525. https://doi.org/10.1016/j.engappai.2010.08.005
7. Mahdavi M, Fesanghary M, Damangir E (2007) An improved harmony search algorithm for solving optimization problems. Appl. Math. Comput. 188:1567–1579. https://doi.org/10.1016/j.amc.2006.11.033
8. Qin AK, Suganthan PN (2005) Self-adaptive differential evolution algorithm for numerical optimization. IEEE Congr. Evol. Comput. 2:1785–1791. https://doi.org/10.1109/CEC.2005.1554904
9. Shi Y, Tan Y, Coello CAC (2014) Advances in swarm intelligence. Springer, Berlin
10. Simon D (2008) Biogeography-based optimization. IEEE Trans. Evol. Comput. 12:702–713. https://doi.org/10.1109/TEVC.2008.919004
11. Storn R, Price K (1997) Differential evolution - a simple and efficient heuristic for global optimization over continuous spaces. J. Glob. Optim. 11:341–359. https://doi.org/10.1023/A:1008202821328
12. Tan Y, Li J, Zheng Z (2014) ICSI 2014 competition on single objective optimization. Technical report, Peking University
13. Tan Y, Li J, Zheng Z (2015) Introduction and ranking results of the ICSI 2014 competition on single objective optimization. arXiv:1501.02128

14. Tan Y, Zhu Y (2010) Fireworks algorithm for optimization. Advances in swarm intelligence, vol 6145. Lecture notes in computer science. Springer, Berlin, pp 355–364. https://doi.org/10.1007/978-3-642-13495-1_44
15. Wang G, Guo L, Duan H, Wang H, Liu L, Shao M (2013) Hybridizing harmony search with biogeography based optimization for global numerical optimization. J. Comput. Theor. Nanosci. 10:2312–2322. https://doi.org/10.1166/jctn.2013.3207
16. Yao X, Liu Y, Lin G (1999) Evolutionary programming made faster. IEEE Trans. Evol. Comput. 3:82–102. https://doi.org/10.1109/4235.771163
17. Zhang B, Zhang M, Zheng YJ (2014) A hybrid biogeography-based optimization and fireworks algorithm. In: Proceedings of IEEE congress on evolutionary computation, pp. 3200–3206. https://doi.org/10.1109/CEC.2014.6900289
18. Zhang B, Zheng YJ, Zhang MX, Chen SY (2017) Fireworks algorithm with enhanced fireworks interaction. IEEE/ACM Trans. Comput. Biol. Bioinform. 14(1):42–55. https://doi.org/10.1109/TCBB.2015.2446487
19. Zheng S, Janecek A, Tan Y (2013) Enhanced fireworks algorithm. In: Proceedings of IEEE congress on evolutionary computation, pp. 2069–2077. https://doi.org/10.1109/CEC.2013.6557813
20. Zheng YJ, Ling HF, Xue JY (2014) Ecogeography-based optimization: Enhancing biogeography-based optimization with ecogeographic barriers and differentiations. Comput. Oper. Res. 50:115–127. https://doi.org/10.1016/j.cor.2014.04.013
21. Zheng YJ, Ling HF, Wu XB, Xue JY (2014) Localized biogeography-based optimization. Soft Comput. 18:2323–2334. https://doi.org/10.1007/s00500-013-1209-1
22. Zheng Y, Wu X (2014) Evaluating a hybrid DE and BBO with self adaptation on icsi 2014 benchmark problems. Advances in swarm intelligence, vol 8795. Lecture notes in computer science. Springer, Berlin, pp 422–433. https://doi.org/10.1007/978-3-319-11897-0_48
23. Zheng YJ, Zhang MX, Zhang B (2014) Biogeographic harmony search for emergency air transportation. Soft Comput. 20:967–977. https://doi.org/10.1007/s00500-014-1556-6

Chapter 6
Application of Biogeography-Based Optimization in Transportation

6.1 Introduction

There are a lot of optimization problems in the field of transportation, some of which can be modeled as continuous optimization problems, while others can be modeled as combinatorial optimization problems. Nowadays, with the development of transportation systems, most of such problems are high-dimensional and/or NP-hard. In recent years, we have adapted BBO algorithm to a variety of transportation problems and achieved good results.

6.2 BBO for General Transportation Planning

Transportation planning is one of the fundamental problems in mathematics and operations research [4, 13]. The problem can be found in roadway transportation, railway transportation, air transportation, etc, but in most cases it focuses on the roadway mode.

6.2.1 A General Transportation Planning Problem

The general transportation planning problem considers there are m supply centers (sources) and n demand targets, and we need to determine the quantity of supplies from each source to each target, such that the overall transportation planning is optimized. There can be one or more types of supplies, and the problem objective can be measured in terms of transportation cost, transportation time, or their combinations. The problem constraints can include the available vehicles, available drivers, the upper limits of cost and time, etc.

© Springer Nature Singapore Pte Ltd. and Science Press, Beijing 2019
Y. Zheng et al., *Biogeography-Based Optimization: Algorithms and Applications*,
https://doi.org/10.1007/978-981-13-2586-1_6

Let us consider a simple version of the problem, where there is only one type of supplies, each target is supplied by only one source, and a source only supplies one target. Let a_i be the available supplies at source i, b_j be the demand of target j, and c_{ij} be the cost for transporting one unit of supplies from source i to target j ($1 \leq i \leq m, 1 \leq j \leq n$), the problem is to determine the quantity x_{ij} of supplies from each source i to each target j, so as to minimize the total transportation cost. The problem can be formulated as follows:

$$\min \quad f = \sum_{i=1}^{m} \sum_{j=1}^{n} c_{ij} x_{ij}$$

$$\text{s.t.} \quad \sum_{i=1}^{m} x_{ij} \geq b_j, \quad j = 1, 2, ..., n \tag{6.1}$$

$$\sum_{j=1}^{n} x_{ij} \leq a_i, \quad i = 1, 2, ..., m$$

$$x_{ij} \geq 0, \quad i = 1, 2, ..., m; \; j = 1, 2, ..., n$$

In practice, a_i and b_j are often limited to positive integers rather than arbitrary real numbers, and thus decision variables x_{ij} are integers. However, the transportation cost are often nonlinear functions of the quantity of supplies. Suppose there is only one type of vehicle with a maximum capacity c, then $\lceil x_{ij}/c \rceil$ vehicles are needed to deliver x_{ij} supplies (where $\lceil \cdot \rceil$ represents rounding up to the nearest integer). Now let c_{ij} denote the transportation cost *per vehicle* from source i to target j, then the transportation cost for x_{ij} supplies is $c_{ij} \lceil x_{ij}/c \rceil$. Let e_i be the number of vehicles available in the ith source, the above transportation planning problem model can be transformed into the following form:

$$\min \quad f = \sum_{i=1}^{m} \sum_{j=1}^{n} c_{ij} \lceil x_{ij}/c \rceil$$

$$\text{s.t.} \quad \sum_{i=1}^{m} x_{ij} \geq b_j, \quad j = 1, 2, ..., n$$

$$\sum_{j=1}^{n} x_{ij} \leq a_i, \quad i = 1, 2, ..., m \tag{6.2}$$

$$\sum_{j=1}^{n} \lceil x_{ij}/c \rceil \leq e_i, \quad i = 1, 2, ..., m$$

$$x_{ij} \in \mathbb{Z}^{+}, \quad i = 1, 2, ..., m; \; j = 1, 2, ..., n;$$

6.2.2 BBO Algorithms for the Problem

We use the Local-BBO and Local-DE/BBO [22] to solve the transportation planning problem (6.2). To handle the constraints of the problem, we define the following penalty functions to calculate the violations of the three constraints for each solution vector $\mathbf{x}$:

$$\psi_j(\mathbf{x}) = \max \left(b_j - \sum_{i=1}^{m} x_{ij}, 0 \right) \tag{6.3}$$

$$\psi_i^1(\mathbf{x}) = \max \left(\sum_{j=1}^{n} x_{ij} - a_i, 0 \right) \tag{6.4}$$

$$\Psi_i^2(\mathbf{x}) = \max \left(\sum_{j=1}^{n} \lceil x_{ij}/c \rceil - e_i, 0 \right) \tag{6.5}$$

and then transform the objective function as follows:

$$\max f'(\mathbf{x}) = -f(\mathbf{x}) - M_1 \left(\sum_{j=1}^{n} \Psi_j(\mathbf{x}) - \sum_{i=1}^{m} \Psi_i^1(\mathbf{x}) \right) - M_2 \sum_{i=1}^{m} \Psi_i^2(\mathbf{x}) \tag{6.6}$$

where we set the coefficients $M_1 = \dfrac{10}{c} \sum_{i=1}^{m} \sum_{j=1}^{n} c_{ij} b_j$ and $M_2 = 10 \sum_{i=1}^{m} \sum_{j=1}^{n} c_{ij} e_i$.

Our algorithms employ the procedure shown in Algorithm 6.1 to initialize each solution $\mathbf{x}$. Note that it cannot always guarantee a feasible solution. When the final rb value is still a positive number, the procedure will be reapplied. However, the procedure can produce solutions in a diverse distribution, and our test indicates that more than 95% solutions are feasible.

Both Local-BBO and Local-DE/BBO adopt the random neighborhood structure for the problem, and their algorithmic frameworks are described in Sects. 3.3 and 5.2, respectively. As the quantity of supplies is often very large, the algorithms simply round any non-integer component to the nearest integer during the evolution.

6.2.3 Computational Experiments

We randomly generate a set of 10 transport planning problem instances, the features of which are shown in Table 6.1. On the test instances, we compare Local-BBO (L-BBO) and Local-DE/BBO (L-DE/BBO) with the following algorithms:

- An exact branch-and-bound (BnB) method for integer programming [13].
- The original BBO algorithm [14].

Algorithm 6.1: Initialize a solution to the general transportation planning problem

---

1 Let $ra = \sum\limits_{i=1}^{m} a_i, rb = \sum\limits_{j=1}^{n} b_j$;   `// available supplies and unsatisfied demands`

2 for $i = 1$ *to* m **do** $ra_i = a_i, re_i = e_i$;

3 for $j = 1$ *to* n **do** $rb_j = b_j$;

4 for $i = 1$ *to* m **do**

5   **for** $j = 1$ *to* n **do**

6   **if** $rb_j = 0$ **then** continue;

7   **if** $ra - ra_i > rb \wedge \mathrm{rand}() < 1 - rb/(ra - ra_i)$ **then** $x_{ij} \leftarrow 0$;

         `// probably do not use i to satisfy j`

8   **else if** $ra_i \geq rb_j \wedge rb_j \leq c$ **then**

9 $x_{ij} \leftarrow rb_j$;   `// one vehicle is sufficient to satisfy j`

10 $ra_i \leftarrow ra_i - rb_j, re_i \leftarrow re_i - 1, rb_j \leftarrow 0, ra \leftarrow ra - rb_j, rb \leftarrow rb - rb_j$;

11   **else if** $ra_i \geq rb_j \wedge rb_j/c \leq re_i \wedge \mathrm{rand}() < i/m$ **then**

12 $x_{ij} \leftarrow rb_j$;   `// probably use source i to satisfy j`

13 $ra_i \leftarrow ra_i - x_{ij}, re_i \leftarrow e_i - \lceil rb_j/c \rceil, rb_j \leftarrow 0, ra \leftarrow ra - rb_j, rb \leftarrow rb - rb_j$;

14   **else**

15 $x_{ij} \leftarrow \mathrm{rand}(c, \min(ra_i, rb_j))$;

16 $ra_i \leftarrow ra_i - x_{ij}, re_i \leftarrow e_i - \lceil x_{ij}/c \rceil, rb_j \leftarrow rb_j - x_{ij}, ra \leftarrow ra - x_{ij}, rb \leftarrow rb - x_{ij}$;

17   **if** $ra_i = 0$ **then** break;

Table 6.1 Summary of the test instances of general transportation planning

No.	m	n	$\sum\limits_{j=1}^{n} b_j$	$\sum\limits_{i=1}^{m} a_i$	$\sum\limits_{i=1}^{m} e_i$
#1	3	5	75	150	15
#2	5	10	173	285	29
#3	6	15	257	533	36
#4	8	16	680	959	112
#5	10	25	793	1622	135
#6	15	30	1058	3303	162
#7	15	50	1985	5151	277
#8	20	50	2219	7130	289
#9	30	80	2620	7669	335
#10	30	100	3855	7353	526

Table 6.2 Running time (in sec) of the algorithms for general transportation planning

No.	BnB	BBO	B-BBO	GA	SGA	DE	DE/BBO	L-BBO	L-DE/BBO
#1	7.3	4.5	4.1	4.6	4.3	3.7	4.2	5.2	4.6
#2	59.2	10.2	9.1	10.4	9.7	8.8	9.8	11.6	10.5
#3	530	33.9	29.5	36.6	30.3	27.7	30.6	36.7	36.3
#4	2763.3	52	46.8	57.5	48.3	41.9	48.9	64.9	59
#5	29736.7	125.4	101.1	140.6	116.7	92.7	118.3	158.3	142.8
#6	256901.3	309.5	280.8	330	288.1	237.9	302.7	365.5	333.1

- The blended BBO (B-BBO) algorithm [11].
- A GA for transportation planning [16].
- An improved GA (denoted as SGA) [6].
- The DE algorithm [15].
- The DE/BBO algorithm [8].

The parameters of the comparative algorithms are set as suggested in the literature. The MNFE is set to $500mn$ for all the heuristic algorithms. On each problem instance, each algorithm is run for 30 times with different random seeds, except that BnB is run once to obtain the exact optimal solution (but on instance #6 it does not terminate after 72 h, and thus its best-known solution is used).

Table 6.2 presents the running time of the BnB method and the average running time of other heuristic algorithms on the first 6 instances. As the instance size increases, the running time of BnB increases dramatically, and the running time of heuristic algorithms becomes much shorter than BnB.

For the ease of comparison, on each of the first six instances, let f° be the exact optimal objective function value obtained by BnB, and on each of the last four instances let f° be the best objective value obtained among the eight heuristic algorithms in all 30 running times, the objective value of the each other algorithm is converted into its ratio to f°. Table 6.3 shows the experimental results, including the best, worst, mean and standard deviation of the resulting objective values of each comparative algorithms over the 30 runs. On each test instance, the best mean values among the eight algorithms are shown in boldface.

As we can see, all the algorithms obtain the optimal solution on the simplest instance #1; on #2 only GA cannot guarantee the optimum; on #3 and #4, GA, SGA and the original BBO cannot guarantee the optima; on the medium-size instance #5, only B-BBO and L-DE/BBO can always obtain the optimum, while the results of GA and SGA are much worse; on larger-size instance #6, no algorithm can obtain the optimal solution within the predefined time (about 5 min), but B-BBO and L-DE/BBO exhibit significant performance advantages over the other algorithms.

In the last four larger-size instances, Local-DE/BBO obtains the best mean value among the eight algorithms, followed by B-BBO, and both of them show much better performance than others. The two GA show the worst performance: On instance #7 and #8, their resulting transportation costs are about 2.5 to 3 times of that of L-DE/BBO, and on #9 and #10, their costs are more than 5 times of L-DE/BBO. The

Table 6.3 Experimental results on general transportation planning

No.	Metrics	BBO	B-BBO	GA	SGA	DE	DE/BBO	L-BBO	L-DE/BBO
#1	Mean	**1.00**	**1.00**	**1.00**	**1.00**	**1.00**	**1.00**	**1.00**	**1.00**
	Best	1.00	1.00	1.00	1.00	1.00	1.00	1.00	1.00
	Worst	1.00	1.00	1.00	1.00	1.00	1.00	1.00	1.00
	Std	0.00	0.00	0.00	0.00	0.00	0.00	0.00	0.00
#2	Mean	**1.00**	**1.00**	1.11	**1.00**	**1.00**	**1.00**	**1.00**	**1.00**
	Best	1.00	1.00	1.00	1.00	1.00	1.00	1.00	1.00
	Worst	1.00	1.00	1.17	1.00	1.00	1.00	1.00	1.00
	Std	0.00	0.00	0.06	0.00	0.00	0.00	0.00	0.00
#3	Mean	1.18	**1.00**	1.24	1.15	**1.00**	**1.00**	**1.00**	**1.00**
	Best	1.00	1.00	1.12	1.00	1.00	1.00	1.00	1.00
	Worst	1.29	1.00	1.33	1.30	1.00	1.00	1.00	1.00
	Std	0.12	0.00	0.10	0.13	0.00	0.00	0.00	0.00
#4	Mean	1.25	**1.00**	1.43	1.29	**1.00**	**1.00**	**1.00**	**1.00**
	Best	1.17	1.00	1.27	1.17	1.00	1.00	1.00	1.00
	Worst	1.38	1.00	1.76	1.50	1.00	1.00	1.00	1.00
	Std	0.11	0.00	0.28	0.20	0.00	0.00	0.00	0.00
#5	Mean	1.30	**1.00**	1.85	1.58	1.12	1.18	1.21	**1.00**
	Best	1.26	1.00	1.63	1.50	1.00	1.12	1.16	1.00
	Worst	1.39	1.00	2.25	1.90	1.18	1.27	1.39	1.00
	Std	0.09	0.00	0.32	0.29	0.09	0.06	0.08	0.00
#6	Mean	1.95	**1.12**	2.26	2.01	1.48	1.36	1.50	1.13
	Best	1.89	1.08	2.06	1.79	1.31	1.22	1.39	1.09
	Worst	2.11	1.13	2.65	2.27	1.68	1.57	1.69	1.15
	Std	0.15	0.02	0.34	0.27	0.17	0.21	0.17	0.03
#7	Mean	2.27	1.07	2.65	2.43	1.35	1.25	1.62	**1.06**
	Best	1.95	1.00	2.21	2.11	1.21	1.13	1.30	1.02
	Worst	2.48	1.13	2.99	2.78	1.58	1.51	1.82	1.12
	Std	0.30	0.08	0.36	0.31	0.19	0.22	0.26	0.06
#8	Mean	2.90	1.09	3.11	3.02	1.33	1.27	1.68	**1.05**
	Best	2.67	1.05	2.98	2.75	1.24	1.18	1.39	1.00
	Worst	3.09	1.14	3.45	3.18	1.55	1.38	1.95	1.13
	Std	0.21	0.05	0.25	0.20	0.15	0.13	0.26	0.05
#9	Mean	4.80	1.11	5.16	4.87	1.93	1.90	2.18	**1.08**
	Best	4.65	1.05	4.85	4.70	1.77	1.73	2.03	1.00
	Worst	5.02	1.20	5.49	5.20	2.10	2.11	2.39	1.16
	Std	0.21	0.08	0.32	0.24	0.18	0.15	0.20	0.06
#10	Mean	5.21	1.16	5.75	5.39	2.24	2.13	2.49	**1.12**
	Best	5.05	1.08	5.51	5.20	2.09	1.98	2.15	1.00
	Worst	5.50	1.26	5.98	5.60	2.35	2.30	2.68	1.21
	Std	0.22	0.11	0.27	0.23	0.13	0.12	0.27	0.09

original BBO performs slightly better than GA. The resulting costs of the remaining three algorithms are about twice of L-DE/BBO. In summary, L-DE/BBO exhibits the best performance, and its advantage is significant on larger-size instances.

6.3 BBO for Emergency Transportation Planning

In this section, we extend the above problem to a typical combinatorial optimization problem, and then design-specific discrete BBO algorithms for the problem [19].

6.3.1 An Emergency Transportation Planning Problem

The extended problem considers transportation under emergency conditions (such as military delivery and disaster relief). Its main features include

- The sources are often large-scale distribution centers with sufficient supplies, so each target needs only one source to meet its supply demands.
- The sources are often far from the targets, but the targets are typically close to each other, so it is reasonable to consider that each source sends a fleet of vehicles to visits its targets one by one.
- The transportation time is more important than cost and thus is used as the objective function to be minimized.

Similarly, we consider there are m sources and n targets, the quantity of supplies with source i is a_i and the quantity of supplies demanded by target j is b_j. We are also given a deadline $\hat{t}_j$ for the arrival of supplies at each target j.

The problem needs to determine not only the quantity of supplies from each source to each target, but also a sequence π_i of targets that the fleet from source i will visit in turn. Let $C(\pi_i)$ be the set of elements in π_i, a solution should satisfy the following constraints:

$$\sum_{j \in C(\pi_i)} b_j \leq a_i, \quad i = 1, 2, \ldots, m \tag{6.7}$$

$$\bigcup_{i=1}^{m} C(\pi_i) = \{1, 2, \ldots, n\} \tag{6.8}$$

$$C(\pi_i) \bigcap C(\pi_{i'}) = \varnothing, \quad \forall i, i' \in \{1, 2, \ldots, m\} \wedge i \neq i' \tag{6.9}$$

Now consider the transportation time. Let $\tilde{t}_{ij}$ be the travel time from source i to target j (which may include the preparation time of source i if needed), $t_{j,j'}$ be the travel time from one target j to another j', and $\text{Index}(\pi_i, j)$ be the index of target j in sequence π_i, then the constraints on arrival time can be expressed as:

$$\tilde{t}_{ij} + \sum_{k=1}^{\text{Index}(\pi_i, j)-1} t_{\pi_i(k), \pi_i(k+1)} \leq \hat{t}_j, \quad j = 1, 2, \ldots, n; \text{Index}(\pi_i, j) > 0 \qquad (6.10)$$

The objective is to minimize the total weighted arrival time (i.e., waiting time) of the targets (where w_j is weight importance of target j):

$$\min f(\mathbf{x}) = \sum_{i=1}^{m} \sum_{j \in C(\pi_i)} w_j \left(\tilde{t}_{ij} + \sum_{k=1}^{\text{Index}(\pi_i, j)-1} t_{\pi_i(k), \pi_i(k+1)} \right) \qquad (6.11)$$

6.3.2   A BBO Algorithm for the Problem

The problem considered here is a typical combinatorial optimization problem, for which we propose a specific discrete BBO algorithm. Each solution to the problem is encoded as a set of m sequences (permutations) $\{\pi_1, \ldots, \pi_i, \ldots, \pi_m\}$, which implicitly satisfies constraints (6.8) and (6.9). When randomly initializing a solution, we first randomly generate a permutation of $\{1, 2, \ldots, n\}$ and then randomly divide it into m parts. If the resulting solution does not satisfy constraint (6.7), it is repaired by the procedure shown in Algorithm 6.2.

Algorithm 6.2: The procedure for repairing a solution to the emergency transportation planning problem

---

1 Divide the set of subsequences into two parts: Π_1 of subsequences that do not satisfy (6.7) and Π_2 of subsequences that satisfy (6.7);

2 For each sequence $\pi_i \in \Pi_1$, let $e_i = \left(\sum_{j \in C(\pi_i)} b_j \right) - a_i$, use the minimal subset sum algorithm [5] to select a subset Δ_i from $C(\pi_i)$ whose total supply demand is minimized under the condition of being larger than e_i;

3 Let Δ be the set of all integers Δ_i;

4 For each sequence $\pi_i \in \Pi_2$, let $c_i = a_i - \left(\sum_{j \in C(\pi_i)} b_j \right)$, and let Γ be the set of all c_i;

5 Using the maximum loading algorithm [20] to load the integers in Δ into the knapsacks with capacities specifies by integers in Γ, and move the corresponding targets from subsequences in Π_1 into subsequences in Π_2.

For constraint (6.10), we use the following penalty function to calculate the constraint violation of each solution $\mathbf{x}$:

$$\Psi_j(\mathbf{x}) = \max \left(\tilde{t}_{ij} + \sum_{k=1}^{\text{Index}(\pi_i, j)-1} t_{\pi_i(k), \pi_i(k+1)} - \hat{t}_j, 0 \right) \qquad (6.12)$$

And thus the objective function is translated into:

$$\max f'(\mathbf{x}) = -f(\mathbf{x}) - M \sum_{j=1}^{n} \Psi_j(\mathbf{x}) \tag{6.13}$$

where M is set to $100 \sum_{j=1}^{n} \hat{t}_j$.

Next we design the migration and mutation operators for the problem. Given two solutions $H = \{\pi_1, \ldots, \pi_i, \ldots, \pi_m\}$ and $H' = \{\pi_1', \ldots, \pi_i', \ldots, \pi_m'\}$; the migration from H to H' at the ith dimension is performed as follows:

(1) Let R_i be the set of tasks in π_i', but not in π_i.
(2) Set $\pi_i' = \pi_i$ in H'.
(3) For each $j \in R_i$, find the position k such that j belongs to π_k of H, and then insert j into π_k' of H', such that the new π_k' has the minimum length of completion time among all possible insertions.

Take an instance with 3 sources and 8 targets as an example, let $H = \{[5, 3], [7, 4], [2, 6, 8, 1]\}$ and $H' = \{[5, 4, 1], [8, 2], [7, 3, 6]\}$; when migrating the first dimension of H to H', we have $R_1 = \{4, 1\}$, and after migration $\pi_1' = [5, 3]$; Target 4 is in π_2 of H and thus will be inserted into $\pi_2' = [8, 2]$, such that π_2' has the minimum length of completion time among all permutations of the set $\{2, 4, 8\}$; similarly, target 1 would be inserted into π_3', such that it has the minimum length of completion time among all permutations of $\{1, 3, 6, 7\}$.

The mutation operation of a solution $H = \{\pi_1, \ldots, \pi_i, \ldots, \pi_m\}$ at the ith dimension is performed by randomly selecting an i' other than i, drop a target j from π_i and reinsert it into $\pi_{i'}$, such that the new $\pi_{i'}$ has the minimum completion time.

If a migration or mutation results in an infeasible solution, Algorithm 6.2 is used to repair the solution. The framework of the discrete BBO for the problem is presented in Algorithm 6.3, where p_m is the mutation rate typically set to 0.02.

6.3.3   Computational Experiments

In experiment, we use 10 test instances generated based on several disaster relief operations in recent years. Table 6.4 summarizes the basic features of the instances, where $\widetilde{t}_{ij}$ is the average transport time (in hours) from a source to a target, $\bar{t}_{j,j'}$ is the average transport time between pairs of targets, and $T^{\max}$ is the upper limit of problem-solving time (in seconds) set according to the needs of emergency response.

For comparison, we have also implemented the following three algorithms:

- A GA based on [2], which uses a two-part chromosome representation: The first part is a permutation of targets, and the second part gives the number of targets assigned to each source. A two-part chromosome crossover operator is used.
- A combinatorial PSO algorithm based on [10], where a particle is an $(m \times n)$-dimensional vector, and each component $x_{ij} \in \{0, 1\}$ is denoted whether target j is assigned to source i.

Algorithm 6.3: The discrete BBO for emergency transportation planning

---

1 Randomly initialize a population P of solutions to the problem;
2 **while** *the stop criterion is not satisfied* **do**
3   Evaluate the fitness of the solutions, based on which calculate their immigration and emigration rates;
4   **for** *each solution $H \in P$* **do**
5   **for** $i = 1$ *to m* **do**
6   **if** rand() $< \lambda(H)$ **then**
7   Select another solution $H' \in P$ with probability $\propto \mu(H')$;
8   Perform migration from H' to H at the ith dimension;

9   Sort the solutions in the decreasing order of fitness;
10   **for** *each solution H in the second half of P* **do**
11   **for** $i = 1$ *to m* **do**
12   **if** rand() $< p_m$ **then**
13   Perform mutation on H at the ith dimension;

14 **return** *the best solution found so far;*

Table 6.4 Summary of the test instances of emergency transportation planning [19]

No.	m	n	$\tilde{t}_{ij}$	$\bar{t}_{j,j'}$	$T^{\max}$
#1	3	10	5.8	0.8	30
#2	5	20	7.9	1.2	30
#3	6	30	9.6	1.0	60
#4	8	40	9.3	1.2	90
#5	10	50	12.6	0.9	120
#6	12	60	16.5	0.9	150
#7	12	75	18.2	0.6	160
#8	15	80	16.8	0.6	180
#9	16	100	21.3	0.5	240
#10	18	118	19.5	0.5	300

- An improved ACO method based on [18], which adds a virtual central depot as the "nest," regards the actual sources as the "entries" to the nest, and regards the targets as "food."

Each algorithm is run for 30 times with different random seeds on each instance. Table 6.5 presents the experimental results, including the *best* and *mean* objective values and the corresponding standard deviations obtained by the algorithms over the 30 runs. All the objective values are scaled to the range of [0,100] for the ease of comparison. We also perform the paired t test between the results of our BBO algorithm and each comparative algorithm, and mark † in columns 2, 5, and 8 if BBO has statistically significant performance improvement over that algorithm (at a confidence level of 95%).

Table 6.5 Experimental results on emergency transportation planning

No.	GA			PSO			ACO			BBO		
	Best	Mean	Std	Best	Mean	Std	Best	Mean	Std	Best	Mean	Std
#1	18.65	18.65	(0.00)	18.65	18.65	(0.00)	18.65	18.65	(0.00)	18.65	18.65	(0.00)
#2	38.93†	39.31	(1.17)	38.93	38.93	(0.00)	38.93	38.93	(0.00)	38.93	38.93	(0.00)
#3	21.90†	26.03	(2.00)	20.93†	21.11	(0.71)	20.93	20.93	(0.00)	20.93	20.93	(0.00)
#4	76.27†	83.89	(8.38)	69.33†	76.40	(4.61)	68.98	69.52	(0.83)	68.98	69.28	(0.35)
#5	30.20†	34.90	(4.21)	27.56†	30.90	(4.12)	25.56	26.96	(1.67)	25.56	26.76	(1.37)
#6	84.81†	91.03	(8.42)	80.27†	86.41	(7.60)	72.35	78.96	(5.02)	71.89	76.81	(5.22)
#7	68.26†	73.72	(8.06)	60.30†	66.67	(7.15)	51.75†	57.97	(4.83)	51.75	55.06	(3.86)
#8	52.15†	57.10	(6.66)	48.91†	53.38	(5.75)	40.27†	46.49	(6.42)	38.66	42.28	(4.96)
#9	83.25†	90.03	(9.73)	77.88†	85.16	(8.06)	72.60†	76.33	(5.85)	70.58	72.27	(5.02)
#10	72.63†	81.01	(10.24)	63.97†	74.16	(8.08)	60.37†	65.05	(6.68)	56.03	58.32	(5.92)

As we can see, on the simplest instance #1, all the algorithms achieve the same optimal value; on instance #2, PSO, ACO and BBO always achieve the same optimum, but GA fails to do so. On the remaining instances, the mean bests of ACO and BBO are always better than GA and PSO. In general, the performance of GA is the worst among the four bio-inspired algorithms; PSO exhibits a relatively fast convergence speed, but it is easier to be trapped by local optima than ACO and BBO. The statistical tests also show that BBO has significant performance advantage over GA and PSO on the 9 and 8 instances, respectively. Comparatively speaking, the search mechanisms of ACO and BBO make them more capable of jumping out of local optima without sacrificing performance, and thus they are more suitable for solving the considered problem.

By comparing ACO and BBO, we find that their performances are similar on small or medium-size instances #1–#6. However, on large-size instances #7–#10, BBO shows performance advantage over ACO, and the advantage becomes more obvious with the increase of the instance size. In summary, the proposed BBO algorithm exhibits the best performance among the four algorithms on the test instances. This also indicates that the BBO metaheuristic has potential in solving a wide range of transportation problems, especially under emergency conditions.

6.4 BBO for Emergency Railway Wagon Scheduling

Compared with road transportation, rail transportation has competitive advantages such as high capacity, less dependence on weather conditions, punctuality in arriving, and this is particularly suitable for large-scale and long-distance freight transportation, especially under emergency situations. However, one of the difficulties in railway transportation planning is that a train contains lots of wagons, and the efficiency of the wagons usage has an important influence on the efficiency of the transportation plan. In [21], we study an emergency railway wagon scheduling problem, and propose a hybrid BBO algorithm for the problem.

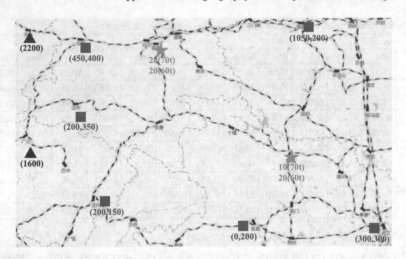

Fig. 6.1 A sample of the Emergency railway wagon scheduling problem. Reprinted from Ref. [21], © 2013, with permission from Elsevier

6.4.1 An Emergency Railway Wagon Scheduling Problem

First we give a simple example of the considered problem on a railway network illustrated by Fig. 6.1. There are two target stations (marked by red triangles) that need to receive 2200 tons and 1600 tons of relief supplies, respectively, and six source stations (blue rectangles) for providing the relief supplies. The train wagons need to be dispatched from two central stations (green stars) to the source stations. There are two types of wagon: The first type has a loading capacity of 60 tons and the second type is of 70 tons. The problem is to allocate the wagons to satisfy the transport demands while delivering the supplies as early as possible. Figure 6.2 presents a solution to the problem instance, where the lines from central stations to source stations are labeled with the number of wagons allocated to the source stations.

In the above case, the network is interconnected, and thus the wagons can be sent from a central station to any source station. However, it is not difficult to extend it to the cases where the network is not completely interconnected.

For this problem, we should carefully allocate wagons not only to satisfy the transport requirements of the source stations, but also to facilitate the fast delivery from the sources to the targets. The problem is very complex not only due to the multiple central, source, and target stations, but also the complex calculation of time whose components include preparation, loading, unloading, travel, waiting, and transfer times.

Formally, the input variables of the problem include

- n: the number of target stations.
- r_j: the quantity of relief supplies required by target station j ($0 < j \leq n$).
- w_j: the weight coefficient of target station j ($0 < j \leq qn$).

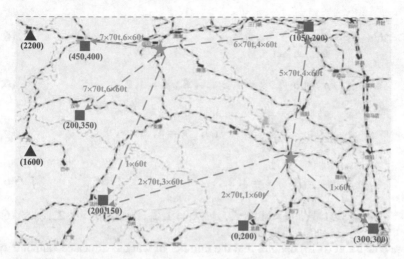

Fig. 6.2 A solution to the sample problem of the railway wagon train. Reprinted from Ref. [21], © 2013, with permission from Elsevier

- m: the number of source stations.
- s_{ij}: the quantity of supplies to be transported from source i to target j ($0 < i \leq m; 0 < j \leq n$).
- d_{ij}: the travel distance between source i and target j ($0 < i \leq m; 0 < j \leq n$).
- K: the number of central stations.
- d_{ik}: the travel distance between source i and center k ($0 < i \leq m; 0 < k \leq K$).
- P: the number of railway wagon types.
- c_p: the carrying capacity of a wagon of type p ($0 < p \leq P$).
- a_{kp}: the available number of wagons of type p at center k ($0 < p \leq P; 0 < k \leq K$).

In some cases, source stations can also provide a part of wagons. When constructing the problem instance, these wagons should be used in prior and the supplies carried by these wagons should be excluded from the problem input.

The decision variables are x_{ikp}, which denotes the number of wagons of type p sent from central station k and to source station i ($0 < i \leq m, 0 < k \leq K, 0 < p \leq P$).

Based on the input and decision variables, we need to calculate some important intermediate variables. The first is y_{ijk}, the quantity of supplies delivered from source i to target j using wagons from center k. The second is the corresponding arrival time t_{ijk} of the supplies. In addition, the wagons are not always fully loaded, and we use Υ_{ij} to denote the redundant wagon capacity for the delivery from source i to target j.

Consequently, the wagon scheduling problem is formulated as follows:

$$\min \ f = \sum_{i=1}^{m} \sum_{j=1}^{n} \sum_{k=1}^{K} w_j y_{ijk} t_{ijk} \tag{6.14}$$

$$\text{s.t.} \ \sum_{i=1}^{m} x_{ikp} \leq a_{kp}, \quad k = 1, 2, \ldots, K; \ p = 1, 2, \ldots, P \tag{6.15}$$

$$\sum_{k=1}^{K} \sum_{p=1}^{P} x_{ikp} \geq \sum_{j=1}^{n} (s_{ij} + \Upsilon_{ij}), \quad i = 1, 2, \ldots, m \tag{6.16}$$

$$\sum_{k=1}^{K} y_{ijk} = s_{ij}, \quad i = 1, 2, \ldots, m; \ j = 1, 2, \ldots, n \tag{6.17}$$

$$x_{ikp} \in \mathbb{Z}^{+}, \quad i = 1, 2, \ldots, m; \ k = 1, 2, \ldots, K; \ p = 1, 2, \ldots, P \tag{6.18}$$

where constraint (6.15) denotes that the number of wagons allocated cannot exceed the number of available wagons, (6.16) denotes that the wagons allocated should provide enough loading capacity to meet the transport requirements, and (6.17) indicates the supply-demand balance.

The above formulation is relatively simple due to the use of the intermediate variables which, however, are not easy to calculate. First consider Υ_{ij}, we expect that the redundant wagon capacity is as small as possible subject to that the loading capacity is enough, and thus its value is determined by the following subproblem:

$$\min \ \Upsilon_{ij} = \left(\sum_{p=1}^{P} x_p c_p \right) - s_{ij} \tag{6.19}$$

$$\text{s.t.} \ \sum_{p=1}^{P} x_p c_p \geq s_{ij} \tag{6.20}$$

$$x_p \in \mathbb{Z}^{+}, \quad p = 1, 2, \ldots, P \tag{6.21}$$

Use the above case study as an example, the loading capacities of the two wagon types are 60 and 70, and the capacity needed by the first source are $s_{11} = 200$, $s_{12} = 150$. Thus, we can obtain that $\Upsilon_{11} = 60 + 2 \times 70 - 200 = 0$, $\Upsilon_{12} = 3 \times 60 - 150 = 30$. Usually, the number of wagon types is small, and hence we can pre-calculate and save all possible combinations of wagons to accelerate the algorithm.

Next consider y_{ijk} and t_{ijk}. For each source i, after the wagons have been allocated to it, we need to distribute the wagons to the targets assigned to the source. The optimization of the distribution scheme can be modeled by the following subproblem:

$$\min \ T_i = \sum_{j=1}^{n} \sum_{k=1}^{K} w_j y_{ijk} t_{ijk} \tag{6.22}$$

$$\text{s.t.} \ \sum_{j=1}^{n} \left(y_{ijk} + \frac{\Upsilon_{ij}}{K} \right) \leq \sum_{p=1}^{P} x_{ikp} c_p, \quad k = 1, 2, \ldots, K \tag{6.23}$$

$$\sum_{k=1}^{K} y_{ijk} = s_{ij}, \quad j = 1, 2, \ldots, n \tag{6.24}$$

$$y_{ijk} \in \mathbb{Z}^{+}, \quad j = 1, 2, \ldots, n; k = 1, 2, \ldots, K \tag{6.25}$$

The constraints (6.23) and (6.24) are derived from the main problem constraints (6.16) and (6.17), respectively. The time t_{ijk} consists of the following parts:

• The travel time from center k to source i, whose value mainly depends on d_{ik}.
• The time for loading supplies, whose value mainly depends on the quantity y_{ijk}.
• The travel time from source i to target j, which depends on both d_{ij} and y_{ijk}.

Thus, t_{ijk} is a function of y_{ijk}, d_{ik}, and d_{ij}, and we use the following empirical equation to calculate it:

$$t_{ijk} = c_1 d_{ik} + c_2 y_{ijk} + c_3 d_{ij} \log_2(c_4 + y_{ijk}) \tag{6.26}$$

where c_1, c_2, c_3, and c_4 are constant coefficients (the value of c_2 usually depends on the loading capability of the station).

The objective function (6.14) of the main problem can be calculated based on m instances of subproblem (6.22):

$$\min T = \sum_{i=1}^{m} T_i \tag{6.27}$$

The wagon scheduling problem is an NP-hard integer programming problem with nonlinear objectives and constraints and multiple subproblems, and thus is suitable to be solved by heuristic algorithms rather than traditional exact algorithms.

6.4.2   A Hybrid BBO/DE Algorithm for the Problem

For solving the problem, we propose a hybrid BBO/DE algorithm with the local neighborhood structure, denoted as HL-BBO/DE [21]. The algorithm first initializes a population of solutions (habitats), and for each solution derives and solves m instances of the subproblem (6.22)–(6.24), the subsolutions of which are then synthesized to evaluate the solution fitness. The solutions are continually improved

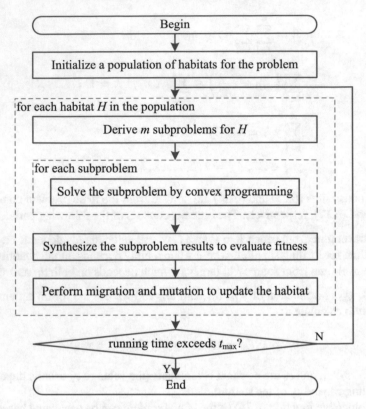

Fig. 6.3 Flowchart of the HL-BBO/DE algorithm. Reprinted from Ref. [21], © 2013, with permission from Elsevier

by migration and mutation until the stop conditions are met. The flowchart of the algorithm is shown in Fig. 6.3.

Each solution to the main problem is an (mKP)-dimensional integer vector. When initializing a solution, each x_{ikp} is set to a random integer in $[0, x_{ijk}^{U}]$ where

$$x_{ijk}^{U} = \min \left[a_{kp}, \frac{\sum_{j=1}^{n} (s_{ij} + \Upsilon_{ij})}{\max_{0 < p \le P} c_p} \right] \tag{6.28}$$

HL-BBO/DE uses the ring neighborhood structure (see Sect. 3.3.1) to maintain the population. It also employs the original BBO migration and the DE/rand/1/bin mutation and then combines the solution H_i' obtained by migration and V_i obtained by mutation as follows (where π_i is the mutation rate of H_i):

$$U_{ij} = \begin{cases} V_{ij}, & \text{rand}() \le \pi_i \\ H_{ij}', & \text{otherwise} \end{cases} \tag{6.29}$$

Any non-integer component generated by mutation is rounded up to the nearest integer. For each solution, the algorithm compares the fitness of H_i', U_i, and the original solution and keeps the best among them in the population. In particular, if both H_i' and U_i are infeasible, the original solution is remained. Moreover, to avoid search stagnation, if a solution keeps unchanged in the population for l^U successive generations, it is replaced by a new solution randomly initialized. If the current optimal solution is not updated after the L^U successive generation, the ring neighborhood structure is reset. The values of parameters l^U and L^U are set according to the size of the problem:

$$l^U = \min(\sqrt[3]{mKP}, 10) \qquad (6.30)$$
$$L^U = \min(\sqrt{mKP}, 20) \qquad (6.31)$$

Algorithm 6.4 presents the framework of HL-BBO/DE for the emergency railway wagon scheduling problem.

Algorithm 6.4: The HL-BBO/DE algorithm for the emergency railway wagon scheduling problem

---

1 Randomly initialize a population of solutions to the problem;
2 **while** *the stop condition is not satisfied* **do**
3   **for** *each H_i in the population* **do**
4   Generate m subproblem instances for H_i;
5   Solve the instances and synthesize the results to evaluate H_i;
6   Calculate the migration and mutation rates of H_i;
7   **for** *each H_i in the population* **do**
8   **for** $d = 1$ *to* mKP **do**
9   **if** rand() $\leq \lambda(H_i)$ **then**
10   Select a neighbor H_j of H_i with probability proportional to $\mu(H_j)$;
11         $H_i(d) \leftarrow H_j(d)$;
12   Perform DE mutation on H_i to generate a vector V_i;
13   Generate a trial vector U_i according to (6.29);
14   Keep best among H_i, the migrated H_i', and U_i in the population;
15   **if** *H_i remains unchanged for l^U generations* **then**
16   Set H_i to a new random solution;
17   Update the best solution H^* found-so-far;
18   **if** *H^* remains unchanged for L^U generations* **then**
19   Reset the ring neighborhood structure;
20 **return** H^*;

Table 6.6 Summary of the test instances of emergency wagon scheduling

No.	m	n	K	$\sum a_{kp}$	$\bar{d}_{ij}$	$\bar{d}_{ik}$	$T^{\max}$
#1	4	1	2	125	469	385	20
#2	6	2	2	150	915	561	40
#3	6	3	4	270	579	602	60
#4	8	3	5	300	1305	390	90
#5	9	3	6	420	1236	515	120
#6	10	4	5	445	735	632	180
#7	12	4	6	450	905	586	240
#8	15	4	8	540	778	605	480
#9	16	5	10	680	712	496	600
#10	18	5	12	720	679	533	900
#11	20	5	14	780	805	677	1200
#12	24	6	15	810	635	520	2400

6.4.3 Computational Experiments

We construct a set of 12 problem instances, which use the three common wagon types with loading capacities of 58t, 60t, and 70t, respectively. Table 6.6 summarizes the features of the instances, where $\sum a_{kp}$ is the total number of available wagons, $\bar{d}_{ij}$ is the average distance between pairs of sources and targets, $\bar{d}_{ik}$ is the average distance between pairs of sources and centers, and $T^{\max}$ is the upper limit of the problem-solving time (in seconds).

We compare HL-BBO/DE with the following algorithms on the instances:

- The BnB algorithm for integer programming [13].
- An ABC algorithm for integer programming [1].
- The DE algorithm [15], where non-integer components are always rounded.
- B-BBO for constraint optimization [11].
- CBBO-DM [9], a BBO constraint optimization algorithm blended with DE.
- Another version of hybrid BBO and DE denoted as H-BBO/DE, which does not use neighborhood structure as HL-BBO/DE.

The ABC and DE are adapted for the considered problem by using the following penalty functions to handle the constraints:

$$\Phi_{ik} = \max \left(\sum_{j=1}^{n} \left(y_{ijk} + \frac{\Upsilon_{ij}}{K} \right) - \sum_{p=1}^{P} (x_{ikp} c_p), 0 \right) \qquad (6.32)$$

$$\Psi_{ij} = \max \left(s_{ij} - \sum_{k=1}^{K} y_{ijk}, 0 \right) \qquad (6.33)$$

Table 6.7 Experimental results of BnB on the first 6 instances. Reprinted from Ref. [21], © 2013, with permission from Elsevier

No.	Running time	f^*
#1	19.5	4.05E+01
#2	45.3	3.40E+03
#3	150.1	7.05E+03
#4	527.2	1.68E+05
#5	1508.2	2.12E+06
#6	6973.5	7.82E+06

and thus transforming the objective function as follows:

$$\min T = \left(\sum_{i=1}^{m} T_i \right) + M \left(\sum_{i=1}^{m} \sum_{k=1}^{K} \Phi_{ik} + \sum_{i=1}^{m} \sum_{j=1}^{n} \Psi_{ij} \right) \tag{6.34}$$

where the coefficient M is set to $c_3 \sum_{i=1}^{m} \sum_{j=1}^{n} d_{ij}$.

The parameters of the comparative algorithms are set as suggested in the literature. First we run BnB to try to find the exact optimal solutions, but on instances #6–#12 BnB cannot stop within four hours. Table 6.7 presents its running time (in seconds) and the resulting objective function values on the first 6 instances, from which we can see that its running time increases very rapidly with the increase of instance size and thus is not suitable for emergency conditions.

The six heuristic algorithms are run for 40 times on each test instance. Table 6.8 presents the results of the algorithms over the 40 runs, where best mean values among the six algorithms are in boldface. As we can see, on instance #1, all the six algorithms obtain the optimal solution; on instance #2, DE and B-BBO fail to do so. In the remaining instances, HL-BBO/DE and H-BBO /DE show significant performance advantages over the others, and HL-BBO/DE always achieves the best result among the algorithms. In more details:

- On small-size #3–#5, the mean best of HL-BBO/DE is about 85–95% of ABC and CBBO-DM and 60–80% of B-BBO and DE.
- On medium-size #6–#8, the mean best of the HL-BBO/DE is about 55–70% of ABC and CBBO-DM and 30–50% of B-BBO and DE.
- On large-size #9–#12, the mean best of the HL-BBO/DE is about 20–40% of ABC and CBBO-DM, and 8–20% of B-BBO and DE.

H-BBO/DE and HL-BBO/DE show similar performance on small and medium-size instances, but on large-scale instances HL-BBO/DE exhibits a significant performance advantage, mainly because its local neighborhood structure can effectively increase solution diversity and suppress premature convergence maturity. In

Table 6.8 Experimental results on emergency wagon scheduling. Reprinted from Ref. [21], Copyright 2013, with permission from Elsevier

No.	Metrics	ABC	DE	B-BBO	CBBO-DM	H-BBO/DE	HL-BBO/DE
#1	Mean	**4.05E+01**	**4.05E+01**	**4.05E+01**	**4.05E+01**	**4.05E+01**	**4.05E+01**
	Best	4.05E+01	4.05E+01	4.05E+01	4.05E+01	4.05E+01	4.05E+01
	Worst	4.05E+01	4.05E+01	4.05E+01	4.05E+01	4.05E+01	4.05E+01
	Std	0	0	0	0	0	0
#2	Mean	**3.40E+03**	3.64E+03	3.68E+03	**3.40E+03**	**3.40E+03**	**3.40E+03**
	Best	3.40E+03	3.40E+03	3.50E+03	3.40E+03	3.40E+03	3.40E+03
	Worst	3.40E+03	3.75E+03	3.73E+03	3.40E+03	3.40E+03	3.40E+03
	Std	0	1.48E+02	9.35E+01	0	0	0
#3	Mean	7.53E+03	7.80E+03	8.45E+03	7.50E+03	7.42E+03	**7.08E+03**
	Best	7.46E+03	7.59E+03	8.18E+03	7.36E+03	7.29E+03	7.05E+03
	Worst	7.62E+03	7.99E+03	8.56E+03	7.63E+03	7.49E+03	7.10E+03
	Std	8.80E+01	3.62E+02	2.59E+02	1.06E+02	7.24E+01	3.16E+01
#4	Mean	2.09E+05	2.32E+05	2.47E+05	1.99E+05	1.96E+05	**1.75E+05**
	Best	1.97E+05	2.19E+05	2.33E+05	1.92E+05	1.89E+05	1.68E+05
	Worst	2.15E+05	2.45E+05	2.55E+05	2.08E+05	2.01E+05	1.79E+05
	Std	9.10E+03	1.63E+04	1.44E+04	1.02E+04	8.65E+03	3.63E+03
#5	Mean	3.28E+06	3.60E+06	3.48E+06	3.24E+06	3.30E+06	**2.90E+06**
	Best	3.21E+06	3.48E+06	3.38E+06	3.16E+06	3.21E+06	2.85E+06
	Worst	3.35E+06	3.73E+06	3.59E+06	3.31E+06	3.37E+06	2.95E+06
	Std	8.75E+04	1.50E+05	1.39E+05	1.11E+05	9.08E+04	4.22E+04
#6	Mean	1.33E+07	1.64E+07	1.75E+07	1.25E+07	1.29E+07	**8.69E+06**
	Best	1.29E+07	1.51E+07	1.72E+07	1.20E+07	1.25E+07	8.55E+06
	Worst	1.35E+07	1.69E+07	1.80E+07	1.27E+07	1.33E+07	8.73E+06
	Std	2.47E+05	5.66E+05	4.09E+05	3.92E+05	2.52E+05	9.47E+04
#7	Mean	9.63E+07	1.35E+08	1.43E+08	9.85E+07	1.06E+08	**6.39E+07**
	Best	9.59E+07	1.26E+08	1.40E+08	9.71E+07	1.03E+08	6.29E+07
	Worst	9.90E+07	1.39E+08	1.48E+08	1.08E+08	1.07E+08	6.46E+07
	Std	1.85E+06	8.55E+06	6.72E+06	2.22E+06	1.73E+06	8.98E+05
#8	Mean	6.94E+09	1.45E+10	1.08E+10	7.27E+09	6.52E+09	**4.00E+09**
	Best	6.79E+09	1.38E+10	1.05E+10	7.12E+09	6.39E+09	3.93E+09
	Worst	7.03E+09	1.48E+10	1.10E+10	7.35E+09	6.61E+09	4.03E+09
	Std	1.15E+08	4.51E+08	2.79E+08	1.39E+08	1.03E+08	5.11E+07
#9	Mean	2.48E+11	5.21E+11	3.98E+11	2.75E+11	2.31E+11	**1.02E+11**
	Best	2.39E+11	5.02E+11	3.80E+11	2.60E+11	2.24E+11	9.78E+10
	Worst	2.51E+11	5.50E+11	4.12E+11	2.83E+11	2.36E+11	1.04E+11
	Std	6.88E+09	2.70E+10	1.79E+10	1.27E+10	5.67E+09	2.04E+09

(continued)

Table 6.8 (continued)

No.	Metrics	ABC	DE	B-BBO	CBBO-DM	H-BBO/DE	HL-BBO/DE
#10	Mean	3.27E+13	1.29E+14	1.00E+14	4.21E+13	2.86E+13	**7.66E+12**
	Best	3.24E+13	1.25E+14	9.96E+13	4.13E+13	2.83E+13	7.58E+12
	Worst	3.30E+13	1.32E+14	1.02E+14	4.26E+13	2.88E+13	7.70E+12
	Std	1.98E+11	3.05E+12	2.65E+12	5.01E+11	1.41E+11	7.79E+10
#11	Mean	9.60E+14	1.90E+15	1.68E+15	9.00E+14	7.16E+14	**1.96E+14**
	Best	9.50E+14	1.86E+15	1.63E+15	8.85E+14	7.10E+14	1.92E+14
	Worst	9.66E+14	1.93E+15	1.70E+15	9.17E+14	7.22E+14	1.99E+14
	Std	9.55E+12	4.18E+13	3.27E+13	2.05E+13	8.85E+12	3.03E+12
#12	Mean	1.45E+15	5.72E+15	3.06E+15	1.64E+15	9.12E+14	**3.53E+14**
	Best	1.43E+15	5.63E+15	3.03E+15	1.62E+15	9.01E+14	3.46E+14
	Worst	1.46E+15	5.80E+15	3.09E+15	1.66E+15	9.16E+14	3.61E+14
	Std	1.52E+13	8.15E+13	2.77E+13	1.90E+13	8.28E+12	5.18E+12

summary, HL-BBO/DE is the most efficient algorithm for the emergency wagon scheduling problem, especially with large number of stations and wagons.

6.5 BBO for Emergency Air Transportation

Air transportation is the most convenient way to deliver supplies, especially when the railways and highways are interrupted in disasters. In [24], we study an emergency air transportation problem and use BHS and EBO [23] to solve it.

6.5.1 An Emergency Air Transportation Problem

We consider that in an emergency event, the affected area needs n types of supplies, the importance weight of supply types j is w_j, and the expected and the lowest amounts of supply type j are r_j and l_j, respectively. The supplies need to be transported by air from a set of m air freight hubs to an airport closest to the affected area. The travel time from hub i to the target airport is t_i, and the amount of supply j available at hub i is s_{ij}. During the operational period, hub i can arrange at most K_i flight batches, where each batch k has a capacity c_{ik} and requires a preparation time τ_k. The problem is to produce a transportation plan, i.e., determine x_{ijk}, the amount of supply j to be delivered in batch k of hub i ($1 \leq i \leq m$; $1 \leq j \leq n$; $1 \leq k \leq K_i$), such that the overall delivery efficiency is as high as possible.

The efficiency of a solution to the problem can be measured in two aspects. The first is the arrival time, and it is easy to see that the expected arrival time of batch k from hub i is $t_{ik} = t_i + \tau_k$. The second is the fulfillment of the supplies. For each supply type j, we define a "fulfillment degree" δ_j as follows:

$$\delta_j = \max \left(r_j, \sum_{i=1}^{m} \sum_{k=1}^{K_i} x_{ijk} \right) - l_j \tag{6.35}$$

The problem objective is to minimize the total weighted arrival time while maximize the total fulfillment degree. By using an "award" coefficient to α balance the two objectives, we formulate the problem as follows:

$$\min \ f(\mathbf{x}) = \sum_{i=1}^{m} \sum_{j=1}^{n} \sum_{k=1}^{K_i} w_j x_{ijk} t_{ik} - \alpha \sum_{j=1}^{n} w_j \delta_j \tag{6.36}$$

$$\text{s.t.} \ \sum_{i=1}^{m} \sum_{k=1}^{K_i} x_{ijk} \geq l_j, \quad j = 1, 2, \ldots, n \tag{6.37}$$

$$\sum_{j=1}^{n} x_{ijk} \leq c_{ik}, \quad i = 1, 2, \ldots, m; k = 1, 2, \ldots, K_i \tag{6.38}$$

$$\sum_{k=1}^{K_i} x_{ijk} \leq s_{ij}, \quad i = 1, 2, \ldots, m; j = 1, 2, \ldots, n \tag{6.39}$$

$$x_{ijk} \in \mathbb{Z}^+, \quad i = 1, 2, \ldots, m; j = 1, 2, \ldots, n; k = 1, 2, \ldots, K_i \tag{6.40}$$

The problem is a multi-type, multi-source, and single-target transportation problem with complex constraints. However, it can be reduced to a much simpler form. It is not difficult to observe that, for each hub i, if we have determined all x_{ij}, in order to optimize the objective function (6.36), the supplies should be sent in decreasing order of their importance weights until the capacities are exhausted, i.e., more important supplies should be arranged to earlier batches. Let t_{ij} be the expected arrival time of supply j from hub i (if x_{ij} is divided into multiple batches, then an arithmetic mean time is used), we can reduce the problem into the following form:

$$\min \ f(\mathbf{x}) = \sum_{i=1}^{m} \sum_{j=1}^{n} w_j x_{ij} t_{ij} - \alpha \sum_{j=1}^{n} w_j \delta_j + M P(\mathbf{x}) \tag{6.41}$$

$$\text{s.t.} \ x_{ij} \leq s_{ij}, \quad i = 1, 2, \ldots, m; j = 1, 2, \ldots, n \tag{6.42}$$

$$x_{ij} \in \mathbb{Z}^+, \quad i = 1, 2, \ldots, m; j = 1, 2, \ldots, n; k = 1, 2, \ldots, K_i \tag{6.43}$$

where M is a large positive constant and $P(\mathbf{x})$ is penalty function for handling the constraints of batch capacities and lowest supply demands:

$$P(j) = \left(\sum_{j=1}^{n} \max \left(l_j - \sum_{i=1}^{m} x_{ij}, 0 \right) \right) + \left(\sum_{i=1}^{m} \max \left(\sum_{j=1}^{n} x_{ij} - \sum_{k=1}^{K_i} c_{ik}, 0 \right) \right) \quad (6.44)$$

Let $\bar{K}$ be the average number of batches of the hubs, now the problem dimension decreases from $mn\bar{K}$ to mn. After solving the reduced problem, by greedily allocating the supplies to the batches, the solution to the original problem is obtained.

6.5.2 BHS and EBO Algorithms for the Problem

The reduced problem is an (mn)-dimensional integer programming problem. Except the input domain and search range, there is no additional constraint, and thus any heuristic algorithm for high-dimensional continuous optimization can be applied to solve the problem. We respectively use EBO (Algorithm 4.1) and BHS (Algorithm 5.6) to solve this problem, where any non-integer component is rounded to the nearest integer.

6.5.3 Computational Experiments

We compare the proposed EBO and BHS algorithm with the following six algorithms (all of which round real-value solution components to the nearest integers):

- The original HS algorithm [7].
- The improved HS (IHS) algorithm [12].
- The differential harmony search (DHS) algorithm [3].
- The original BBO algorithm [14].
- The B-BBO algorithm [11].
- Another hybrid HS and BBO (HSBBO) algorithm [17], which differs from BHS in that it uses the basic clonal-based migration.

The test set consists of 8 problem instances constructed from real-world disaster relief operations in China in recent years, including 2008 Wenchuan earthquake (08W), 2010 Yushu earthquake (10Y), 2010 Zhouqu mudslides (10Z), 2011 Yingjiang earthquake (11Y), 2012 Ninglang earthquake (12N), 2013 Yaan earthquake (13Y), 2013 Dingxi earthquake (13D), and 2014 Yingjiang earthquake (14Y) [24]. Table 6.9 summarizes the basic features of the instances, where $\bar{K}$, $\bar{t}$, and $\bar{r}$ are the average number of batches, average travel time (in hours) from the sources to the target, and the average amount of supplies (in tons), respectively, and $T^{\max}$ is the upper limit of the running time (in seconds).

Table 6.9 Summary of the test instances of emergency air transportation

No.	Event	m	n	$\bar{K}$	$\bar{t}$	$\bar{r}$	$T^{\max}$
#1	12N	5	8	4.8	5.4	96	300
#2	14Y	8	8	8.5	7.1	132	300
#3	11Y	9	7	12.5	6.2	106	300
#4	13D	12	9	10.3	6.5	65	480
#5	10Z	16	6	16.0	5.0	98	480
#6	10Y	15	10	18.7	6.0	84	600
#7	13Y	19	9	15.5	4.6	109	600
#8	08W	28	18	23.5	4.8	75	900

Table 6.10 presents the experimental results on the instances, where the best object values among the eight algorithms are in boldface. The BHS and EBO algorithms exhibit much better performance than the others: BHS obtains the best results on six instances, while EBO does so on four instances. In details, five algorithms (except BBO, HS, and DHS) obtain the same optimal solution on instance #1, and three algorithms including HIS, BHS, and EBO obtain the same optimal solution on instance #2; on the remaining instances, the results of BHS and/or EBO are always the best among the eight algorithms; the larger the instance size, the more obvious the performance advantage of BHS and EBO over the other algorithms. The results indicate that the hybrid BHS is much more effective than the individual BBO and HS, while the new migration mechanism of EBO is as efficient as the hybrid BBO migration and harmony search.

Comparing BHS with EBO, BHS performs better than EBO on instances #3–#5 and #7, while EBO performs better on #6 and #8. That is because BHS integrating HS with blended migration has more local search ability, while EBO has more global search ability in exploring large solution spaces of instances #6 and #8 with a large number of flight batches.

6.6   Summary

This chapter describes applications of BBO and its improved versions for transportation problems. For the transportation planning problem, we design discrete migration and mutation operators. For the other three problems, we use integer- or real-valued encoding such that the original BBO operators can be directly applied. In next chapter, we will use more specific discrete operators for job scheduling.

Table 6.10 Experimental results on emergency air transportation

No.	Metrics	BBO	B-BBO	HS	HIS	DHS	HSBBO	BHS	EBO
#1	Mean	1.01E+06	**8.21E+05**	8.58E+05	**8.21E+05**	8.32E+05	**8.21E+05**	**8.21E+05**	**8.21E+05**
	Std	3.05E+05	0.00E+00	6.19E+04	0.00E+00	1.56E+04	0.00E+00	0.00E+00	0.00E+00
	Best	8.21E+05	8.21E+05	8.21E+05	8.21E+05	8.21E+05	8.21E+05	8.21E+05	8.21E+05
	Worst	1.33E+06	8.21E+05	9.27E+05	8.21E+05	8.50E+05	8.21E+05	8.21E+05	8.21E+05
#2	Mean	9.01E+06	7.69E+06	9.29E+06	**7.05E+06**	7.29E+06	7.18E+06	**7.05E+06**	**7.05E+06**
	Std	1.39E+06	3.17E+05	1.50E+06	0.00E+00	3.02E+05	2.32E+05	0.00E+00	0.00E+00
	Best	7.62E+06	7.35E+06	7.35E+06	7.05E+06	7.05E+06	7.05E+06	7.05E+06	7.05E+06
	Worst	9.98E+06	8.11E+06	1.16E+07	7.05E+06	7.75E+06	7.59E+06	7.05E+06	7.05E+06
#3	Mean	9.56E+06	7.27E+06	1.08E+07	7.20E+06	7.98E+06	7.18E+06	**6.52E+06**	6.59E+06
	Std	2.36E+05	8.33E+05	3.16E+07	1.86E+06	1.05E+06	9.33E+05	0.00E+00	9.24E+04
	Best	9.26E+06	6.52E+06	8.85E+06	6.52E+06	6.52E+06	6.52E+06	6.52E+06	6.52E+06
	Worst	9.80E+06	8.07E+06	1.49E+07	9.26E+06	9.03E+06	9.26E+06	6.52E+06	6.76E+06
#4	Mean	5.09E+07	5.35E+07	5.79E+07	3.81E+07	4.25E+07	4.06E+07	**3.63E+07**	3.69E+07
	Std	7.34E+06	2.27E+06	6.08E+06	1.21E+06	5.00E+06	2.69E+06	3.99E+05	1.35E+06
	Best	3.89E+07	4.90E+07	5.24E+07	3.67E+07	3.67E+07	3.69E+07	3.60E+07	3.60E+07
	Worst	5.77E+07	5.50E+07	6.13E+07	3.97E+07	4.63E+07	4.20E+07	3.69E+07	3.80E+07
#5	Mean	1.78E+07	2.20E+07	3.45E+07	1.63E+07	1.95E+07	1.89E+07	**1.36E+07**	1.50E+07
	Std	2.08E+06	2.76E+06	1.03E+07	3.47E+06	3.00E+06	2.75E+06	8.59E+05	1.51E+06
	Best	1.69E+07	1.69E+07	1.84E+07	1.25E+07	1.69E+07	1.54E+07	1.25E+07	1.33E+07
	Worst	1.95E+07	2.38E+07	5.08E+07	1.90E+07	2.56E+07	2.16E+07	1.39E+07	1.68E+07
#6	Mean	2.33E+08	1.85E+08	1.26E+100	8.21E+07	1.17E+08	2.06E+08	6.13E+07	**5.82E+07**
	Std	4.85E+07	2.66E+07	7.92E+99	1.68E+07	9.52E+06	9.21E+07	3.25E+06	3.89E+06
	Best	2.02E+08	1.57E+08	2.02E+08	5.15E+07	1.05E+08	8.99E+07	5.15E+07	5.22E+07
	Worst	3.01E+08	2.36E+08	3.05E+100	9.27E+07	1.29E+08	5.11E+08	6.39E+07	6.08E+07
#7	Mean	3.26E+09	1.80E+09	3.09E+78	1.78E+09	1.76E+09	1.76E+09	**1.73E+09**	1.75E+09
	Std	5.17E+08	7.12E+07	4.68E+78	6.15E+07	7.08E+07	1.85E+07	3.03E+07	3.85E+07
	Best	3.03E+09	1.76E+09	2.15E+09	1.74E+09	1.72E+09	1.73E+09	1.72E+09	1.72E+09
	Worst	3.76E+09	1.88E+09	3.31E+79	1.82E+09	1.82E+09	1.78E+09	1.76E+09	1.76E+09
#8	Mean	7.36E+98	2.05E+12	1.12E+101	1.95E+11	2.57E+12	1.80E+12	6.60E+10	**5.93E+10**
	Std	8.31E+98	2.89E+12	3.67E+101	1.33E+11	3.00E+12	8.94E+11	2.21E+10	8.55E+09
	Best	1.60E+12	6.93E+11	8.01E+12	1.32E+11	9.24E+11	2.76E+11	3.86E+10	3.86E+10
	Worst	9.80E+99	7.79E+12	1.06E+102	5.05E+11	7.03E+12	3.05E+12	8.62E+10	6.75E+10

References

1. Akay B, Karaboga D (2009) Solving integer programming problems by using artificial bee colony algorithm. In: AI*IA, pp 355–364. https://doi.org/10.1007/978-3-642-10291-2-36
2. Carter AE, Ragsdale CT (2006) A new approach to solving the multiple traveling salesperson problem using genetic algorithms. Eur J Oper Res 175:246–257. https://doi.org/10.1016/j.ejor.2005.04.027
3. Chakraborty P, Roy GG, Das S, Jain D, Abraham A (2009) An improved harmony search algorithm with differential mutation operator. Fundam Inform 95:401–426. https://doi.org/10.3233/FI-2009-157

4. Charnes A, Cooper WW (1954) The stepping stone method of explaining linear programming calculations in transportation problems. Manag Sci 1:49–69. https://doi.org/10.1287/mnsc.1.1.49
5. Cormen TH, Leiserson CE, Rivest RL, Stein C (2009) Introduction to algorithms, 3rd edn. MIT Press, Cambridge
6. Fogel DB (1994) An introduction to simulated evolutionary optimization. IEEE Trans Neural Netw 5:3–14. https://doi.org/10.1109/72.265956
7. Geem ZW, Kim JH, Loganathan G (2001) A new heuristic optimization algorithm: harmony search. Simulation 76:60–68. https://doi.org/10.1177/003754970107600201
8. Gong W, Cai Z, Ling CX (2010) DE/BBO: a hybrid differential evolution with biogeography-based optimization for global numerical optimization. Soft Comput 15:645–665. https://doi.org/10.1007/s00500-010-0591-1
9. Ilhem B, Amitava C, Patrick S, Mohamed AN (2012) Biogeography-based optimization for constrained optimization problems. Comput Oper Res 39:3293–3304. https://doi.org/10.1016/j.cor.2012.04.012
10. Jarboui B, Damak N, Siarry P, Rebai A (2008) A combinatorial particle swarm optimization for solving multi-mode resource-constrained project scheduling problems. Appl Math Comput 195:299–308. https://doi.org/10.1016/j.amc.2007.04.096
11. Ma H, Simon D (2011) Blended biogeography-based optimization for constrained optimization. Eng Appl Artif Intel **24**, 517–525. https://doi.org/10.1016/j.engappai.2010.08.005
12. Mahdavi M, Fesanghary M, Damangir E (2007) An improved harmony search algorithm for solving optimization problems. Appl Math Comput 188:1567–1579. https://doi.org/10.1016/j.amc.2006.11.033
13. Rardin R (1998) Optimization in operations research. Pearson Education
14. Simon D (2008) Biogeography-based optimization. IEEE Trans Evol Comput 12:702–713. https://doi.org/10.1109/TEVC.2008.919004
15. Storn R, Price K (1997) Differential evolution - a simple and efficient heuristic for global optimization over continuous spaces. J Global Optim 11:341–359. https://doi.org/10.1023/A:1008202821328
16. Vignaux G, Michalewicz Z (1991) A genetic algorithm for the linear transportation problem. IEEE Trans Syst Man Cybern 21:445–452. https://doi.org/10.1109/21.87092
17. Wang G, Guo L, Duan H, Wang H, Liu L, Shao M (2013) Hybridizing harmony search with biogeography based optimization for global numerical optimization. J Comput Theor Nanosci 10:2312–2322. https://doi.org/10.1166/jctn.2013.3207
18. Yu B, Yang ZZ, Xie JX (2011) A parallel improved ant colony optimization for multi-depot vehicle routing problem. J Oper Res Soc 62:183–188. https://doi.org/10.1057/jors.2009.161
19. Zhang MX, Zhang B, Zheng YJ (2014) Bio-inspired meta-heuristics for emergency transportation problems. Algorithms 7:15–31. https://doi.org/10.3390/a7010015
20. Zheng Y, Shi H, Chen S (2012) Algorithm design. Science Press
21. Zheng YJ, Ling HF, Shi HH, Chen HS, Chen SY (2014) Emergency railway wagon scheduling by hybrid biogeography-based optimization. Comput Oper Res 43:1–8. https://doi.org/10.1016/j.cor.2013.09.002
22. Zheng YJ, Ling HF, Wu XB, Xue JY (2014) Localized biogeography-based optimization. Soft Comput 18:2323–2334. https://doi.org/10.1007/s00500-013-1209-1
23. Zheng YJ, Ling HF, Xue JY (2014) Ecogeography-based optimization: enhancing biogeography-based optimization with ecogeographic barriers and differentiations. Comput Oper Res 50:115–127. https://doi.org/10.1016/j.cor.2014.04.013
24. Zheng YJ, Zhang MX, Zhang B (2014) Biogeographic harmony search for emergency air transportation. Soft Comput 20:967–977. https://doi.org/10.1007/s00500-014-1556-6

Chapter 7
Application of Biogeography-Based Optimization in Job Scheduling

Abstract Job scheduling problems are a class of combinatorial optimization problems that occur in many areas including production, maintenance, education. In this chapter, we adapt BBO to solve a set of scheduling problems including flow-shop scheduling, job-shop scheduling, maintenance job scheduling, and university course timetabling. The experimental results demonstrate the effectiveness and efficiency of BBO for the problems.

7.1 Introduction

Job scheduling problems are a class of well-known combinatorial optimization problems that occur in many practical applications including production scheduling, maintenance job scheduling, course timetabling. This chapter introduces our recent work on the adaptation of BBO for a set of scheduling problems, the key of which is to design effective migration and mutation operators for the discrete solution spaces of the problems.

7.2 BBO for Flow-Shop Scheduling

7.2.1 Flow-Shop Scheduling Problem

Flow-shop scheduling problem (FSP) [26] is to schedule n independent jobs on m machines. Each job contains exactly m operations. The jth operation of each job must be executed on the jth machine ($1 \leq j \leq m$). No machine can perform more than one operation simultaneously. For each operation of each job, processing time is specified. Because the processing times of different jobs on machines are often

© Springer Nature Singapore Pte Ltd. and Science Press, Beijing 2019   143
Y. Zheng et al., *Biogeography-Based Optimization: Algorithms and Applications*,
https://doi.org/10.1007/978-981-13-2586-1_7

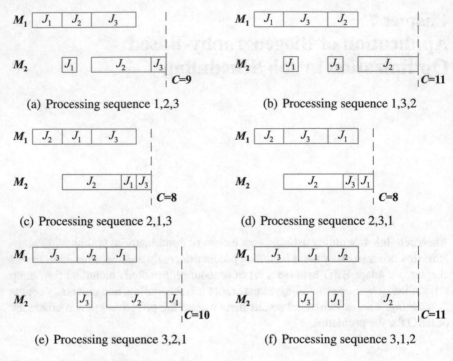

Fig. 7.1 An example of FSP

different, FSP needs to determine a sequence of jobs so as to minimize the completion time of last job (also called makespan of the sequence).

Take a problem with three jobs and two machines as an example: suppose the processing time of job J_1 is 2 on machine M_1 and is 1 on machine M_2, the processing times of the job J_2 on the two machines are respectively 2 and 4, and the processing times of the job J_3 on the two machines are respectively 3 and 1. Figure 7.1 presents all possible six job execution sequences (also known as Gantt charts), among which sequences {2, 1, 3} and {2, 3, 1} have the minimal makespan of 8.

Many problems in other areas can also be expressed in the form of FSP, such as determining the sequence of a set of courses attended by a team of students so as to minimize the completion time of the last student, determining the sequence of stations to be passed by a group of trains so as to minimize the arrival time of last train, and so on.

Formally, the problem input variables include:

- $\{J_1, J_2, ..., J_n\}$, a set of n jobs.
- $\{M_1, M_2, ..., M_m\}$, set of m machines.
- t_{ij}, the processing time of job J_i by machine M_i ($1 \leq i \leq n, 1 \leq j \leq m$).

The problem needs to decide a permutation $\pi = \{\pi_1, \pi_2, ..., \pi_n\}$ of the n jobs. Let $C(\pi_i, j)$ denote the completion time of job π_i on machine M_j. For the first machine

M_1, we have

$$C(\pi_1, 1) = t_{\pi_1, 1} \tag{7.1}$$

$$C(\pi_i, 1) = C(\pi_{i-1}, 1) + t_{\pi_i, 1}, \quad i = 2, ..., n \tag{7.2}$$

The first job π_1 can be processed on machine M_j immediately after it is completed on machine M_{j-1}:

$$C(\pi_1, j) = C(\pi_1, j-1) + t_{\pi_1, j}, \quad j = 2, ..., m \tag{7.3}$$

Each subsequent job π_i can be processed on machine M_j only when the following two conditions are satisfied: (1) the job π_i has been completed on machine M_{j-1}; (2) its previous job π_{i-1} has been completed on machine M_j. So we have

$$C(\pi_i, j) = \max\{C(\pi_{i-1}, j), C(\pi_i, j-1)\} + t_{\pi_i, j}, \quad i = 2, ..., n; j = 2, ..., m \tag{7.4}$$

Thus, the makespan of π is

$$C_{\max}(\pi) = C(\pi_n, m) \tag{7.5}$$

The goal of the problem is to find an optimal π^* in the set Π of all possible sequences to minimize the makespan:

$$C_{\max}(\pi^*) = \min_{\pi \in \Pi} C_{\max}(\pi) \tag{7.6}$$

FSP is similar to TSP in that both of them aim to find an optimal sequence, but the objective function of FSP is much more complex than TSP and thus incurs a higher requirement for the search ability of the problem-solving algorithm.

The basic FSP only requires that each solution is a permutation of n jobs, but there can be some additional constraints, such as posing a due time $d(\pi_i)$ for each job J_i

$$C(\pi_i, m) \le d(\pi_i), \quad i = 1, 2, ..., n \tag{7.7}$$

Using the penalty function method, the objective function of the problem is transformed to

$$f(\pi) = \min\left(C_{\max}(\pi) + M \sum_{i=1}^{n} (C(\pi_i, m) - d(\pi_i), 0) \right) \tag{7.8}$$

It has been proved that FSP is NP-hard when $n \ge 3$ [10]. In practice, n and m are often very large, and thus, traditional methods are inefficient to tackle with combinational explosion. In recent decades, heuristic algorithms have become popular approaches for solving FSP.

7.2.2 A BBO Algorithm for FSP

Solution encoding and initialization The most direct way to encode a solution to FSP is a permutation of jobs, for which special heuristic operators are needed [8, 13]. Another encoding scheme is based on the largest ranked value (LRV) representation, where each solution is a real-number vector, and the jobs are sequenced according to the ranks of the vector components [38]. For example, considering a solution vector [0.72, 0.65, 0.33, 0.95, 0.23, 0.83], the largest value 0.95 is at the fourth dimension, so job 4 is firstly processed; the second largest value 0.83 is at the sixth dimension, so job 6 is secondly processed; by analogy, we can get the job sequence $\pi = [4, 6, 1, 2, 3, 5]$. Using this scheme, heuristic algorithms such as BBO and DE for high-dimensional continuous optimization problems can be directly used for FSP.

Here, we apply BBO to solve FSP. As we know, meta-heuristics like GA and BBO are general-purpose algorithms that rarely use specific properties of the problem, while traditional problem-specific methods utilize such properties. One typical method is the Nawaz–Enscore–Ham (NEH) method [28] that uses a greedy strategy to construct a potential good solution to FSP, the procedure of which is shown in Algorithm 7.1.

Algorithm 7.1: The NEH procedure [28]

---

1 For each job J_i, compute $T_i = \left(\sum_{j=1}^{n} t_{ij} \right)$, the total processing time on all machines;
2 Sort the job set in increasing order of T_i;
3 Let $k = 2$, take the first two jobs to construct a permutation π that minimizes the makespan;
4 **while** $k < n$ **do**
5   Let $k = k + 1$, take the kth job, temporarily insert it into all possible k positions in π, and finally select the position which has the minimum makespan to insert the job;

6 **return** π;

NEH cannot guarantee to find the optimal solution to FSP. Experiments show that NEH has good performance on some small-size instances, but does not work well on large-size instances. Nevertheless, the solution obtained by NEH has some useful information that can contribute to the search of other algorithms [8].

Our BBO algorithm for FSP employs NEH in population initiation. That is, we first use NEH to generate a job sequence π and then randomly initialize a population of N solutions (real-value vectors), among which randomly select a solution worse than π and replace it with π.

Mutation Using the LRV-based real-number encoding, the original BBO mutation is directly applicable. To enhance solution diversity, we further define two additional mutation operators, one exchanging two components randomly selected, and the other reversing a subsequence randomly selected. The three mutations are called general mutation, exchanging mutation, and reverse mutation, respectively.

Enhanced local search Researches have shown that performing local exploitation around some promising solutions can greatly facilitate the evolution of population-based heuristic algorithms [18, 23]. We design a reinsertion local search operator based on the job with longest waiting time (JLWT). The waiting time τ_{ij} of job π_i on machine M_j is

$$\tau_{ij} = \min\left(C(\pi_{i-1}, j) - C(\pi_i, j-1), 0\right) \tag{7.9}$$

And the total waiting time T_i of job π_i on all machines is

$$T_i = \sum_{j=2}^{m} \tau_{ij} \tag{7.10}$$

The local search operator takes the job with maximum T_i from the permutation, temporarily reinserts it into the other $(n-1)$ positions, and replaces the solution with the best one among the $(n-1)$ solutions if it is better than the original solution.
The algorithm framework Algorithm 7.2 presents the proposed BBO algorithm for FSP.

Algorithm 7.2: The BBO algorithm for FSP

---

1 Randomly initialize a population of solutions, including a solution produced by NEH;
2 **while** *the stop condition is not satisfied* **do**
3   Calculate the migration and mutation rates of the solutions;
4   **foreach** *solution* **x** *in the population* **do**
5   Perform BBO migration on **x** to produce a new solution **x'**;
6   **if** $f(\mathbf{x'}) < f(\mathbf{x})$ **then**
7   Replace **x** with **x'** in the population;
8   **if** **x** *is better than the current best* **x*** **then**
9                 $\mathbf{x^*} \leftarrow \mathbf{x}$;
10   Perform JLWT-based local search on **x***;
11   **else**
12   Randomly select one of the general, exchanging, and reverse mutations;
13   Perform the mutation on **x'**;

14 **return x***;

7.2.3 Computational Experiments

We respectively use the original BBO [33], B-BBO [24], Local-DE/BBO using random neighborhood structure (LDB) [42], and EBO [43] to implement the above framework , and compare them with the following three popular algorithms for FSP:

- Another hybrid BBO algorithm [38], denoted by HBBO.
- A PSO-based memetic algorithm [23], denoted by PSOMA.
- A hybrid DE algorithm [31], denoted by HDE.

We select seven FSP test instances from [32], whose size and known best makespan C^* are given in columns 2 and 3 of Table 7.1. For the first four BBO algorithms, we set the same parameters as follows: the population size $N = 50$, the maximum migration rates $I = E = 1$, and the maximum mutation rate $\pi_{max} = 0.06$. The parameters of HBBO, PSOMA, and HDE are set as suggested in the literature.

On each instance, we run each algorithm for 30 times with different random seeds and set the same stop condition as that NFE reaches 100000. We record the resulting best, worst, and average makespan, denoted by C_{best}, C_{worst}, and C_{avg}, respectively. For ease of comparison, we use the following three metrics to normalize the makespan:

- BRE: the best relative error with C^*.

$$\text{BRE} = (C_{best} - C^*)/C^* \times 100\% \qquad (7.11)$$

- ARE: the average relative error with C^*.

$$\text{ARE} = (C_{avg} - C^*)/(C^*) \times 100\% \qquad (7.12)$$

- WRE: the worst relative error with C^*.

$$\text{WRE} = (C_{worst} - C^*)/C^* \times 100\% \qquad (7.13)$$

Table 7.1 presents the experimental results of the comparative methods on the test instances (the results of the three comparative methods are taken from [38], and the missing data is marked by "–").

As we can see, all the BBO algorithms achieve better results than PSOMA and HDE, which indicates that BBO heuristics are effective for FSP. Among the four versions of our BBO, only the original BBO performs worse than HBBO, mainly because its clonal-based migration limits the solution diversity. The other three versions perform much better than HBBO, which demonstrates the effectiveness of the proposed migration, mutation, and local search operators.

Among the seven algorithms, the overall performance of EBO is the best. In particular, B-BBO, LDE, and EBO can obtain the optimal makespan on the first five instances (the BRE values are 0), and only LDE and EBO can do so on the sixth instances. This indicates that the proposed mutation operator can effectively improve population diversity in very large solution spaces, and the local search operator enables extensive exploitation around promising solutions to reach the optimum.

Table 7.1 Experimental results on test instances of FSP

Instance	$n \times m$	C^*	Metric	PSOMA	HDE	BBO	HBBO	B-BBO	LDB	EBO
Rec01	20×5	1247	BRE	0	0	0	0	0	0	0
			ARE	0.144	0.152	0.031	0.016	0	0	0
			WRE	0.160	–	0.052	0.160	0	0	0
Rec07	20×10	1566	BRE	0	0	0	0	0	0	0
			ARE	0.986	0.920	0.731	0	0.322	0.125	0
			WRE	1.149	–	0.980	0	0.859	0.813	0
Rec13	20×15	1930	BRE	0.259	0.259	0	0	0	0	0
			ARE	0.893	0.705	0.509	0.238	0.196	0.167	0
			WRE	1.502	–	1.502	0.985	1.263	0.985	0
Rec19	30×10	2093	BRE	0.43	0.287	0.287	0.287	0	0	0
			ARE	1.313	0.908	0.926	0.401	0.381	0.275	0.126
			WRE	2.102	–	1.753	0.860	0.860	0.860	0.287
Rec25	30×15	2513	BRE	0.835	0.676	0	0	0	0	0
			ARE	2.085	1.429	0.733	0.541	0.480	0.318	0.133
			WRE	3.233	3.233	1.154	1.154	1.154	0.676	0.541
Rec31	50×10	3045	BRE	1.51	0.427	0.427	0.263	0.263	0	0
			ARE	2.254	1.192	1.035	0.450	0.399	0.287	0.163
			WRE	2.692	–	1.510	1.149	1.149	0.427	0.263
Rec37	75×20	4951	BRE	2.101	1.697	1.697	1.353	1.353	1.133	1.133
			ARE	3.537	2.632	2.367	1.759	1.653	1.517	1.368
			WRE	4.039	–	3.860	2.262	2.101	2.101	1.697

7.3 BBO for Job-Shop Scheduling

7.3.1 Job-Shop Scheduling Problem

Similar to FSP, job-shop scheduling problem (JSP) is also to assign n jobs to m machines, where each job has m operations, each of which to be processed by a machine. The difference is that in JSP different jobs can have different processing sequences. For example, in a freighter loading and unloading workshop, an arrived freighter has a sequence of box hanging, box opening, cargo unloading, and box sealing, while a departing freighter has a sequence of box opening, cargo loading, box sealing, and box hanging.

Therefore, besides the job set, the machine set, and the matrix $T = \left[t_{ij}\right]_{(n \times m)}$ of job-machine processing times (here t_{ij} denotes the processing time of the jth operation of job J_i rather than the time of job J_i on machine M_j, the input to JSP also contains a processing order matrix $S = \left[s_{ij}\right]_{(n \times m)}$, where each ith row is the processing sequence of job J_i. The problem is to determine an execution sequence of jobs for each machine such that the makespan $C_{\max}$ of all jobs is minimized.

Fig. 7.2 Illustration of a JSP instance

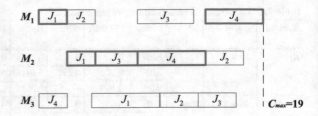

The decision variables can be represented as a matrix $X = \left[x_{ij}\right]_{(m \times n)}$, where x_{jk} represents the kth job executed on machine M_j.

Consider a JSP instance with four jobs and three machines, and the processing order matrix S and time matrix T are as follows:

$$S = \begin{bmatrix} 1 & 2 & 3 \\ 1 & 3 & 2 \\ 2 & 1 & 3 \\ 3 & 2 & 1 \end{bmatrix}, \quad T = \begin{bmatrix} 2 & 2 & 5 \\ 2 & 3 & 5 \\ 3 & 5 & 3 \\ 3 & 6 & 6 \end{bmatrix}$$

A solution to the instance can be represented by a (3×4)-dimensional matrix. For the following solution X, its job scheduling is shown in Fig. 7.2, from which we can get its makespan is $C_{\max} = 19$.

$$X = \begin{bmatrix} 1 & 2 & 3 & 4 \\ 1 & 3 & 4 & 2 \\ 4 & 1 & 2 & 3 \end{bmatrix}$$

Let $C(j, k)$ denote the completion time of kth job on machine M_j, $i = x_{j,k}$ be the corresponding job index, and p be the index of M_j in the processing sequence of J_i. When $k = 1$ and $p = 1$, J_i is the first job on M_j; when $k = 1$ and $p > 1$, J_i is processed after the previous $p - 1$ jobs have been processed on M_j:

$$C(j, 1) = \begin{cases} t_{i,1}, & p = 1 \\ C(s_{i,p-1}, i) + t_{i,p}, & p \neq 1 \end{cases}, \quad j = 1, 2, \ldots, m \tag{7.14}$$

When $k > 1$, the job can be processed by M_j only after its previous $(k - 1)$ operations have been processed:

$$C(j, k) = \begin{cases} C(j, k - 1) + t_{i,p}, & p = 1 \\ \max\left(C(j, k - 1), C(s_{i,p-1}, i)\right) + t_{i,p}, & p \neq 1 \end{cases}, \quad j = 1, 2, \ldots, m; k = 2, 3, \ldots, n$$

$$\tag{7.15}$$

And thus, the makespan is

$$C_{\max}(X) = \max_{1 \le j \le m} C(j, n) \qquad (7.16)$$

The problem objective is to find a X^* that has minimum makespan among the set Π of all possible schedules:

$$C_{\max}(X^*) = \min_{X \in \Pi} C_{\max}(X) \qquad (7.17)$$

Given n jobs and m machines, the number of all possible solutions of FSP is $n!$ while that of JSP is $(n!)^m$. So JSP is much more complex than FSP [19].

7.3.2 An Enhanced BBO Algorithm for the Problem

We propose a hybrid BBO and local search algorithm for JSP. As the term "memetic algorithm" is used in the community of optimization to refer an evolutionary algorithm enhanced by local search [27], we call the algorithm as the biogeography-based memetic optimization algorithm (BBMO).

Migration BBMO encodes each solution to JSP as a $(m \times n)$-dimensional matrix and regards each row (i.e., the job sequence on each machine) as an SIV, is treated as a whole. When the jth SIV of a solution X (denoted by $X(j)$) is immigrated from another solution X', BBMO replaces a random subsequence in $X(j)$ with the corresponding subsequence in $X'(j)$ and makes substitution on the remaining components of $X(j)$ to keep it as a permutation of all jobs $\{J_1, \ldots, J_n\}$.

Considering a JSP instance with six jobs, when performing migration from $X'(j) = [1, 3, 5, 6, 4, 2]$ to $X(j) = [1, 2, 3, 4, 5, 6]$, if the subsequence $[3, 5, 6]$ is selected from X'_j, the migration is performed as follows:

(1) Replace 2 in $X(j)$ with the first element 3 of the subsequence, and change the original 3 in $X(j)$ to 2.
(2) Replace 2 in $X(j)$ with the second element 5 of the subsequence, and change the original 5 in $X(j)$ to 2.
(3) Replace 4 in $X(j)$ with the third element 6 of the subsequence, and change the original 6 in $X(j)$ into 4.

The process is shown in Fig. 7.3, and after migration $X(j) = [1, 3, 5, 6, 2, 4]$.

Algorithm 7.3 presents the migration procedure of BBMO.

Mutation The BBMO mutation operator is also performed on an SIV (a job sequence on a machine) at a time, by swapping two randomly selected jobs in the sequence. Algorithm 7.4 presents the mutation procedure of BBMO.

Fig. 7.3 An example of
BBMO migration

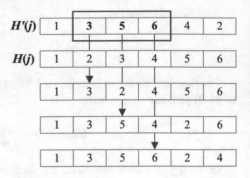

Algorithm 7.3: BBMO migration on a solution X

1 **for** $j = 1$ *to* n **do**
2   **if** rand() $< \lambda(X)$ **then**
3   Select another X' with a probability in proportional to $\mu(X')$;
4   Let $p_1 = $ rand$(1, n/2)$, $p_2 = $ rand$(p_1 + 1, n)$; //subsequence from p_1 to p_2
5   **for** $p = p_1$ *to* p_2 **do**
6   Let $p' = $ Indexof$(X(j), X(j)(p))$;
7        $X(j)(p') \leftarrow X(j)(p)$;
8        $X(j)(p) \leftarrow X'(j)(p)$;

Algorithm 7.4: BBMO mutation on a solution X

1 **for** $j = 1$ *to* n **do**
2   **if** rand() $< \pi(X)$ **then**
3   Let $p_1 = $ rand$(1, n)$;
4   Let $p_2 = $ If $(p_1 < n)$ rand$(p_1 + 1, n)$ else rand$(1, n - 1)$;
5      $X(j)(p_1) \leftrightarrow X(j)(p_2)$;

Local search Whenever BBMO finds a new best solution X^*, it performs a local
search around the solution based on "critical path" method which involves the following concepts [4]:

- *Critical path*: The longest path from the start node to the end node, and its length
  is $C_{\max}$.
- *Critical operation*: The operations on the critical path.
- *Critical block*: A subsequence that has the maximum number of successive critical
  operations on a machine.

Fig. 7.4 Changing operations in the critical block on M_2 in Fig. 7.2

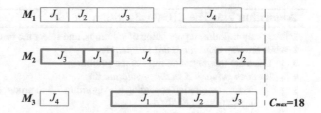

For example, the five operations marked by bold rectangular in Fig. 7.2 are critical operations, and they constitute a critical path. J_1 and J_4 are the two critical blocks on machine M_1, $\{J_1, J_3, J_4\}$ is a critical block on machine M_2, and there is no critical block on M_3.

A critical block has two good properties that can contribute to local search [4]:

- Swapping any two jobs in a critical block will not affect the feasibility of scheduling.
- Swapping any two inner operations (except the first and the last operations) will not affect the makespan.

Thus, we can limit the local search to the change of a critical block, using one of the following three ways [30]:

- Swapping the first and the last operations.
- Moving the an inner operation to the first or the last position.
- Moving the first or the last operation into the block.

For example, for the critical block $\{J_1, J_3, J_4\}$ on machine M_2 in Fig. 7.2, if we move J_3 to the first position, the makespan of the new schedule is reduced to 18, as shown in Fig. 7.4.

The local search operator of BBMO uses the last two ways as they have been shown to be more effective than the first one. Supposing that a solution has B critical blocks and the average length of each block is L, the number of the neighborhood solutions is $2B(L - 2)$.

The algorithm framework Algorithm 7.5 presents the framework of the BBMO for JSP.

7.3.3 Computational Experiments

We respectively use the original BBO [33], RingBBO, and RandBBO [42] to implement the above algorithmic framework, and compare them with the following four popular algorithms for JSP:

Algorithm 7.5: The BBMO algorithm for JSP

---

1 Randomly initialize a population of solutions, and select the best as X^*;
2 **while** *the stop criterion is not satisfied* **do**
3 | Calculate the migration rates of the solutions;
4 | **for** *each solution X in the population* **do**
5 | | Perform migration according to Algorithm 7.3 to produce a new X';
6 | | **if** $f(X') < f(X)$ **then**
7 | | | Replace X with X';
8 | | | **if** $f(X) < f(X^*)$ **then** perform local search on X;

9 | **for** *each solution X in the worst half of the population* **do**
10 | | Perform mutation according to Algorithm 7.4 to produce a new X';
11 | | **if** $f(X') < f(X)$ **then**
12 | | | Replace X with X';
13 | | | **if** $f(X) < f(X^*)$ **then** perform local search on X;

14 **return** X^*;

- A hybrid GA with local search [12], denoted by HGA.
- A hybrid PSO and artificial immune system (AIS) algorithm [11], denoted by HPA.
- A hybrid PSO, SA, and local search algorithm [21], denoted by MPSO.
- A "best-so-far" ABC algorithm [2].

The parameters of BBMO are set as follows: the population size $N = 2mn$, the maximum migration rates $I = E = 1$, the mutation rate $\pi = 0.02$. The parameters of the other four algorithms are set as suggested in the literature. The maximum number of iterations of all algorithms is set to 1000.

We select 40 JSP instances from [20, 36], and the sizes and the best known makespans C^* are given in columns 2 and 3 of Table 7.2. On each instance, we run each algorithm 20 times and present the BRE values of the algorithms in Table 7.2.

As we can see, compared with the other four algorithms, BBMO shows competitive performance. In terms of average BRE value, RandBBMO performs the best, and RingBBMO and BBMO are ranked third and fifth, respectively. Among the 40 instances, BBMO and RingBBMO, respectively, obtain the best known solution on 33 and 36 instances, much better than the HGA, HIA, and MPSO; RandBBMO and ABC obtain the best known solution on 38 instances, and the average BRE value of RandBBMO is less than ABC.

Comparing the three BBMO algorithms, both RingBBMO and RandBBMO have significant performance improvement over BBMO, which shows that neighborhood structure can effectively suppress premature convergence. The overall performance of RandBBMO is slightly better than RingBBMO, which shows that dynamic random neighborhood structure has more exploitation ability.

Table 7.2 Results of comparative experiments on test instances of JSP

Name	$n \times m$	C^*	HGA	HPA	MPSO	ABC	BBMO	RingBBMO	RandBBMO
LA01	10×5	666	0	0	0	0	0	0	0
LA02	10×5	655	0	0	0	0	0	0	0
LA03	10×5	597	0	0	0	0	0	0	0
LA04	10×5	590	0	0	0	0	0	0	0
LA05	10×5	593	0	0	0	0	0	0	0
LA06	15×5	926	0	0	0	0	0	0	0
LA07	15×5	890	0	0	0	0	0	0	0
LA08	15×5	863	0	0	0	0	0	0	0
LA09	15×5	951	0	0	0	0	0	0	0
LA10	15×5	958	0	0	0	0	0	0	0
LA11	20×5	1222	0	0	0	0	0	0	0
LA12	20×5	1039	0	0	0	0	0	0	0
LA13	20×5	1150	0	0	0	0	0	0	0
LA14	20×5	1292	0	0	0	0	0	0	0
LA15	20×5	1207	0	0	0	0	0	0	0
LA16	10×10	945	0	0	0	0	0	0	0
LA17	10×10	784	0	0	0	0	0	0	0
LA18	10×10	848	0	0	0	0	0	0	0
LA19	10×10	842	0	0	0	0	0	0	0
LA20	10×10	902	0.554	0	0	0	0	0	0
LA21	15×10	1046	0	0	0	0	0	0	0
LA22	15×10	927	0.863	0.539	0.539	0	0.863	0	0
LA23	15×10	1032	0	0	0	0	0	0	0
LA24	15×10	935	1.925	0.642	1.604	0	1.604	0	0
LA25	15×10	977	0.921	0	0.205	0	0	0	0
LA26	20×10	1218	0	0	0	0	0	0	0
LA27	20×10	1235	1.700	0.324	1.700	0	0	0	0
LA28	20×10	1216	1.316	0	0.905	0	0	0	0
LA29	20×10	1152	3.819	1.823	2.778	1.042	1.823	1.042	0
LA30	20×10	1355	0	0	0	0	0	0	0
LA31	30×10	1748	0	0	0	0	0	0	0
LA32	30×10	1850	0	0	0	0	0	0	0
LA33	30×10	1719	0	0	0	0	0	0	0
LA34	30×10	1721	0	0	0	0	0	0	0
LA35	30×10	1888	0	0	0	0	0	0	0
LA36	15×15	1268	0.868	0.789	1.025	0	0.868	0.789	0.789
LA37	15×15	1397	0.787	1.002	1.288	0	1.002	0	0
LA38	15×15	1196	1.923	1.003	1.421	0	1.003	1.003	0
LA39	15×15	1233	1.054	0	1.054	0	0	0	0
LA40	15×15	1222	1.555	0.245	1.473	0.164	0.245	0.245	0.245
Average			0.432	0.159	0.350	0.030	0.193	0.077	0.026

7.4　BBO for Maintenance Job Assignment and Scheduling

7.4.1　A Maintenance Job Assignment and Scheduling Problem

The problem model Here, we consider a problem that first assigns n maintenance jobs to m teams and then schedules the jobs of each team. Compared with FSP and JSP, this problem has the following additional challenges:

- It integrates assignment and scheduling of the jobs.
- Different teams and jobs are located at different places, and the time for teams moving between the locations should be taken into consideration.
- The processing times and moving times are imprecise and thus represented in the form of fuzzy variables, due to the uncertainty of the environment [34].
- Unlike machines, human teams need to have rest time during the operation.
- Different jobs have different importance weights. The weights are typically given by experts, but multiple experts may have different opinions.

Besides the number n of jobs and m of teams, the problem has the following input variables:

- $\widetilde{t}_{ij}$, the time for team i to complete job j.
- $\overrightarrow{t}^{\,i}_{j}$, the time for team i to move from its original location to the location of job j.
- $\overrightarrow{t}^{\,i}_{jj'}$, time for team i to move from the location of job j to the location of job j'.
- $\bar{t}_i$, the preparation time of team i before it can set off to work.
- ΔT_i, the upper limit of the continuous working time of team i.
- $\widehat{t}_i$, the rest time needed by team i if its continuous working time reaches the upper limit.
- C_i, the set of jobs that can be carried by team i.
- K, the numbers of experts.
- $\mathbf{w}_k = [w_{k1}, \ldots, w_{kj}, \ldots, w_{kn}]$, the importance weights given by expert k.

Among the above variables, $\widetilde{t}_{ij}$, $\overrightarrow{t}^{\,i}_{j}$, and $\overrightarrow{t}^{\,i}_{jj'}$ are given in the form of fuzzy numbers (such as interval or triangular fuzzy numbers).

Each solution to the problem is a set of m sequences $\{\pi_1, \ldots, \pi_i, \ldots, \pi_m\}$, where π_i is the schedule of jobs assigned to team i. Let $\text{Set}(\pi)$ denote the set of elements in sequence π, and two basic constraints of the problems are as follows:

$$\text{Set}(\pi_i) \subseteq C_i, \quad i = 1, 2, \ldots, m \tag{7.18}$$

$$\bigcup_{i=1}^{m} \text{Set}(\pi_i) = \{1, 2, \ldots, n\} \tag{7.19}$$

Based on the input and decision variables, we can calculate the expected completion time τ_j of each job j, and the problem objective is to minimize the total weighted completion time of all jobs:

$$\min f = \sum_{j=1}^{n} w_j \tau_j \tag{7.20}$$

where w_j is the aggregated weight of job j. Next, we describe our approach for calculating exact completion time from fuzzy time variables and aggregation of weights given by different experts.

The defuzzification method We use the fuzzy expected value model [22] to defuzzificate the fuzzy numbers. Given a fuzzy number $\tilde{\xi}$, its expected value, α-optimistic value, and α-pessimistic value are, respectively, calculated as follows:

$$E(\tilde{\xi}) = \int_0^{+\infty} C_r\{\tilde{\xi} \geq r\}\, dr - \int_{-\infty}^0 C_r\{\tilde{\xi} \leq r\}\, dr \tag{7.21}$$

$$O_\alpha(\tilde{\xi}) = \sup\{r \,|\, C_r\{\tilde{\xi} \geq r\} \geq \alpha\} \tag{7.22}$$

$$P_\alpha(\tilde{\xi}) = \inf\{r \,|\, C_r\{\tilde{\xi} \geq r\} \geq \alpha\} \tag{7.23}$$

where α is a confidence value in $(0,1]$, and C_r is a credibility function satisfying normality, monotonicity, self-duality, and maximality axioms [22]. The advantages of the above measures include low computational complexity and no special requirements on ranking preference.

Based on the above model, for an interval fuzzy number $\tilde{\xi} = (\xi_l, \xi_u)$, we have

$$E(\tilde{\xi}) = (\xi_u + \xi_l)/2 \tag{7.24}$$

$$O_\alpha(\tilde{\xi}) = \begin{cases} \xi_l, & \alpha \leq 0.5 \\ \xi_u, & \alpha > 0.5 \end{cases} \tag{7.25}$$

$$P_\alpha(\tilde{\xi}) = \begin{cases} \xi_u, & \alpha \leq 0.5 \\ \xi_l, & \alpha > 0.5 \end{cases} \tag{7.26}$$

And for a triangular fuzzy number $\tilde{\xi} = (\xi_u, \xi_m, \xi_l)$, we have

$$E(\tilde{\xi}) = (\xi_l + 2\xi_m + \xi_u)/4 \tag{7.27}$$

$$O_\alpha(\tilde{\xi}) = \begin{cases} 2\alpha\xi_m + (1 - 2\alpha)\xi_u, & \alpha \leq 0.5 \\ (2\alpha - 1)\xi_l + (2 - 2\alpha)\xi_m, & \alpha > 0.5 \end{cases} \tag{7.28}$$

$$P_\alpha(\tilde{\xi}) = \begin{cases} (1 - 2\alpha)\xi_l + 2\alpha\xi_m, & \alpha \leq 0.5 \\ (2 - 2\alpha)\xi_m + (2\alpha - 1)\xi_u, & \alpha > 0.5 \end{cases} \tag{7.29}$$

For the considered problem with fuzzy inputs, we first set the confidence level α and then defuzzificate the fuzzy numbers as follows:

(1) For a fuzzy processing time $\tilde{t}_{ij}$, use its α-pessimistic value for a hard job j and use the expected value otherwise.
(2) For a fuzzy travel time $\overrightarrow{t}^{\,i}_{\,j}$ or $\overrightarrow{t}^{\,i}_{\,jj'}$, use its expected value.
(3) Calculate the sum of two fuzzy components based on arithmetic operations on fuzzy numbers [6]. If the α-optimistic value of accumulated continuous working time reaches the limit ΔT^u_i, we consider that team i needs to take a rest.

Multi-expert weight aggregation Usually, experts with differing opinions can give different weights to the jobs, and such differing opinions should be taken into consideration before in a final solution. Moreover, we often need a set of solutions, including a recommended solution and several candidate solutions, to the decision-maker. There are different methods for multi-expert weight aggregation [25], and here we employ an multi-objective approach based on two criteria to enable searching for a diverse set of solutions. The first criterion uses the average weighted sum of completion time:

$$f_1 = \sum_{j=1}^{n} \overline{w}_j \tau_j \qquad (7.30)$$

where $\overline{w}_j$ is the average weight of job j ($1 \leq j \leq n$):

$$\overline{w}_j = \frac{1}{K} \sum_{k=1}^{K} w_{kj} \qquad (7.31)$$

The second criterion always chooses a $\mathbf{w}_k$ which maximizes weighted sum of completion time:

$$f_2 = \max_{1 \leq k \leq K} \left(\sum_{j=1}^{n} w_{kj} \tau_j \right) \qquad (7.32)$$

and the result of which can minimize the dissatisfaction or "cost of disagreement" among all experts [16].

In this way, we get a bi-objective optimization problem and can search for a set of Pareto-optimal solutions to the problem. In practice, we can recommend the solutions in increasing order of f_2 values , while the incorporation of f_1 can not only improve solution diversity but also suppress premature convergence [17].

7.4.2 A Multi-objective BBO Algorithm for the Problem

For solving the above bi-objective problem, we propose a multi-objective BBO algorithm [44]. As there are two objectives, we employ the fast non-dominated sorting method in NSGA-II [5] to calculate a non-dominated ranking K_H for each solution in the population based on the number of solutions it dominates, and then calculate its migration rates as follows (where N_P is the number of non-dominated fronts of the population:

$$\lambda_H = I \frac{k_H}{N_P} \tag{7.33}$$

$$\mu_H = E\left(1 - \frac{k_H}{N_P}\right) \tag{7.34}$$

Let two solutions be $H = \{\pi_1, \ldots, \pi_i, \ldots, \pi_m\}$ and $H' = \{\pi'_1, \ldots, \pi'_i, \ldots, \pi'_m\}$, the migration from H' to H' at the ith is shown in Algorithm 7.6.

Algorithm 7.6: The BBO migration procedure for maintenance job assigning and scheduling.

---

1 Let R_i be the set of jobs in π_i but not in π'_i;
2 Set $\pi_i = \pi'_i$ in H;
3 **foreach** $j \in R_i$ **do**
4   Find the position k such that j belongs to π'_k of H';
5   Insert j into π_k of H such that the new π_k has the minimum length of completion time among all possible insertions;

For example, suppose there are three teams and nine jobs, $H = \{[3,2,6],[7,9,5], [4,1,8]\}$, $H' = \{[2,5],[4,1,3,7],[6,9,8]\}$, when migrating the first dimension of H' to H, we have $R_1 = \{3, 6\}$ and set $\pi_1 = [2, 5]$; job 3 is in π'_2 of H', and thus it will be inserted into $\pi_2 = [7, 9, 5]$ such that π_2 has the minimum completing time among all permutations of $\{3, 7, 9, 5\}$; similarly, job 6 will be inserted into π_3.

In particular, if a migration operation results in an infeasible solution, the operation is skipped on the current dimension and moved to the next dimension.

The mutation operation on a solution $H = \{\pi_1, \ldots, \pi_i, \ldots, \pi_m\}$ is as follows: Randomly select two indices i and i' such that $C_i \cap C_{i'} \neq \emptyset$, drop a job $j \in C_i \cap C_{i'}$ from π_i, and reinsert it into $\pi_{i'}$ such that the new $\pi_{i'}$ has the minimum completion time.

As most multi-objective optimization algorithms, our algorithm uses an archive A for maintaining non-dominated solutions found so far. We also set an upper limit $|A|^u$ of the archive size to avoid performance degradation: When $|A|$ reaches $|A|^u$, a new non-dominated solution H will replace an archive one only when the archive diversity can be potentially improved by doing so. Here, we use the minimum pairwise distance, i.e., the Euclidean distance between two solutions (namely H_a and H_b) which is the minimum among all pairs in A, for the diversity measure [14, 40]:

- If $\min_{H' \in A \wedge H' \neq H_a} \text{dis}(H, H') > \text{dis}(H_a, H_b)$, then replace H_a with H.
- If $\min_{H' \in A \wedge H' \neq H_b} \text{dis}(H, H') > \text{dis}(H_a, H_b)$, then replace H_b with H.
- Otherwise, choose from A closest H_c to H: If $\min_{H' \in A \wedge H' \neq H_c} \text{dis}(H, H') > \text{dis}(H, H_c)$, then replace H_c with H, otherwise discard H.

Algorithm 7.7 presents the framework of the multi-objective BBO for the maintenance job assigning and scheduling problem.

Algorithm 7.7: The multi-objective BBO for maintenance job assigning and scheduling

---

1 Randomly initialize a population P of solutions to the problem;
2 Create an empty archive A;
3 **while** *the stop criterion is not satisfied* **do**
4 | Perform fast nondominated sorting on P;
5 | Calculate the migration rates of the solutions based on Eqs. (7.33) and (7.34);
6 | Update A with the first nondominated front of P;
7 | **foreach** $H \in P$ **do**
8 | | **for** $i = 1$ *to* m **do**
9 | | | **if** rand$() < \lambda_h$ **then**
10 | | | | Select another solution $H' \in P$ with probability $\propto \mu_{H'}$;
11 | | | | migrate from H' to H at the ith dimension according to Algorithm 7.6;
   |
12 | **foreach** $H \in P$ *with* $n_H > 0$ **do**
13 | | Mutate H to a new solution H';
14 | | **if** H' *is not dominated by* H **then** replace H with H';

15 **return** A;

7.4.3 Computational Experiments

Based on data from the disaster operations of 2008 Wenchuan earthquake (08W), 2010 Yushu earthquake (10Y), 2010 Zhouqu mudslides (10Z), 2011 Yingjiang earthquake (11Y), 2012 Ninglang earthquake (12N), the 2013 Ya'an earthquake (13Y), we respectively construct six instances of the emergency maintenance job assigning and scheduling problem, the summary of which is presented in Table 7.3, where MRT is the maximum running time (in minutes) set according to the emergency requirements of the operations.

For each instance, a team of 5–6 experts have assigned importance weights to the jobs with the help of an interactive decision support system. For comparison, we also implement the following four algorithms which have shown good performance on similar scheduling problems:

- A tabu search (TS) algorithm that starts from an arbitrary solution and repeatedly moves from the current solution to the best of neighboring solutions that is not tabu [41]. The tabu tenure is set to $(mn/20)^{1/4}$, and the maximum number of non-improving iterations is set to $mn/2$.
- A GA that uses a two-chromosome encoding with crossover and mutation operators [3]. The population size is 100, and the mutation probability is 0.05.

Table 7.3 Summary of the instances of maintenance job assigning and scheduling. Reproduced from Ref. [44] by permission of John Wiley & Sons Ltd

No.	m	n	$\sum_i \sum_j t_{ij}$	Fuzzy items	MRT
08W	58	626	108525	5813	30
10Y	15	67	3997	95	15
10Z	16	39	3920	93	15
11Y	12	55	2285	106	15
12N	9	32	562	63	15
13Y	21	102	6335	278	20

- A hybrid GA that utilizes path-relinking and TS (denoted by HGAT) [35]. The population size is 60, the mutation probability is 0.2, the tabu tenure is 8, and the maximum number of non-improving iterations is 5.
- Another hybrid GA with simulated annealing (denoted by HGAS) previously used by the organization. The population size is 60, the mutation probability is 0.05, and the cool efficient is 0.96.

We also establish a simplified model for ex post evaluation of a solution to the problem. Let T be the total length of time of the relief operation, τ_j be the completion time of job j in the solution, then the actual effectiveness of the solution is evaluated as

$$E = \sum_{j=1}^{n} (T - \tau_j) e_j \qquad (7.35)$$

where e_j denotes the effectiveness achieved by completing job j. For a ruin excavation job, e_j is calculated as the number of victims released; for a roadway repair job, e_j is the traffic flow passing through the repaired point during the operation; for a structure reinforcement job, e_j is the potential casualties (estimated by experts) maybe caused by the collapse of the structures. The larger the value of E, the better solution.

On each instance, we run each algorithm for 30 times with different random seeds and compute the mean bests of f_2 (Eq. (7.32)) and E (Eq. (7.35)) of the recommended solutions averaged over the 30 runs. Table 7.4 presents the experimental results (scaled to the range of [0,100], the best values are in boldface).

As we can see, our BBO algorithm exhibits the best performance on all the instances. The more complex the instances, the more obvious the performance advantages of BBO. On the simplest instance 12N, the effectiveness value evaluated on the recommended solution of BBO is about 148.26% of that of GA, 122.07% of that of HGAT, and 114.04% of that of HGAS. 08W and 13Y are much more complex instances. On 08W, the effectiveness value achieved by BBO is about 340.66% of GA, 243.84% of HGAT, and 218.09% of HGAS; on 13Y, the effectiveness value of BBO is about 286.05%, 189.16%, and 161.36% of GA, HGAT, and HGAS, respectively.

Table 7.4 Experimental results on instances of maintenance job assigning and scheduling. Reproduced from Ref. [44] by permission of John Wiley & Sons Ltd

Problem	Metric	min f_2					max E				
		TS	GA	HGAT	HGAS	BBO	TS	GA	HGAT	HGAS	BBO
08W	Mean	72.27	25.80	19	17.88	**9.55**	9.19	27.35	38.21	42.72	**93.17**
	Std	20.66	6.30	4.24	3.92	1.95	2.67	8.50	9.10	10.94	16.85
10Y	Mean	88.75	30.16	27.11	28.33	**23.96**	8.90	23.08	29.67	28.85	**36.36**
	Std	23.99	5.98	4.57	7.03	3.25	2.13	5.14	6.27	5.77	5.81
10Z	Mean	65.67	28.68	22.22	20.49	**15.87**	4.29	13.66	24.09	27.20	**46.55**
	Std	15.33	7.82	3.98	5.04	2.73	1.29	3.69	6.50	6.89	8.54
11Y	Mean	90.33	72.30	67.22	64.93	**50.12**	5.62	9.48	13.01	13.86	**20.67**
	Std	15.06	9.57	7.32	10.28	7.20	0.88	1.59	1.53	2.77	3.84
12N	Mean	56.26	50.18	46.15	41.75	**39.90**	11.06	17.20	20.89	22.36	**25.50**
	Std	10.61	6.75	5.35	8.68	5.68	1.72	2.62	2.94	4.28	3.60
13Y	Mean	59.53	47.25	39.67	36.75	**25.83**	18.14	26.67	40.33	47.28	**76.29**
	Std	17.20	9.44	6.52	9.12	5.16	4.59	6.10	7.11	11.06	13.86

Such a remarkable improvement can contribute greatly to reinforce the relief operation and mitigate the damage.

TS algorithm has the worst performance among the five algorithms, mainly because it performs a single-thread search and does not simultaneously evolve a population of solutions. Among the four population-based evolutionary algorithms, GA is easily trapped in local optima and thus exhibits the worst performance. Enhanced by local search procedures, HGAT and HGAS perform better than GA, and there is no statistically significant difference between them. In summary, the results demonstrate that the migration and mutation operators designed for the considered problem can achieve a good balance between exploration and exploitation and thus make our BBO algorithm very efficient in searching the solution space.

We can also find that the value of E is generally inversely proportional to the value of f_2, which indicates that a solution minimizing the objective function is very likely to maximize the effectiveness.

Furthermore, we compare the multi-objective BBO with a single-objective BBO version that solely optimizes f_2. The results show that the multi-objective BBO can always produce Pareto-optimal solutions which are better than the results of the single-objective version. Figure 7.5 presents the convergence curves of the two versions of the six instance, which show that the single-objective BBO typically converges faster at early search stages, but later is easily trapped by local optima and thus is outperformed by the multi-objective BBO. This demonstrates that the multi-objective algorithm can effectively escape from local optima and produce much better solutions.

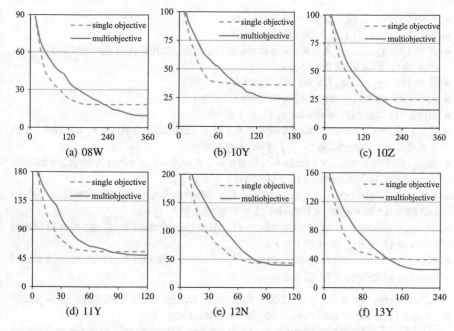

Fig. 7.5 Convergence curves of the single-objective BBO and multi-objective BBO on the maintenance job assigning and scheduling instances. Reproduced from Ref. [44] by permission of John Wiley & Sons Ltd

7.5 BBO for University Course Timetabling

7.5.1 A University Course Timetabling Problem

University course timetabling problem (UCTP) is a typical timetabling problem that has been proven to be NP-hard [9]. It is to schedule a number of university course events over a prefixed period of time (normally a week) while satisfying a variety of soft and hard constraints on rooms, timeslots, students, teachers, etc. Timetabling needs to be done at each semester. In a university, the number of courses, teachers, and students are much more than that in a primary or middle school, and thus, traditional timetabling methods are often time consuming and often result in inefficient timetables that would significantly increase the cost and difficulty of implementations. Nowadays, heuristic optimization algorithms have been popular ways for producing high-quality timetables that can much improve the efficiency of both teachers and students.

Formally, the problem has the following input variables:

- $E = \{e_1, e_2, \ldots, e_n\}$, a set of n events of courses, including lectures, speeches, laboratories.
- $R = \{r_1, r_2, \ldots, r_m\}$, a set of m rooms.

- $T = \{t_1, t_2, \ldots, t_o\}$, a set of o timeslots.
- $S = \{s_1, s_2, \ldots, s_p\}$, a set of p students.
- $F = \{f_1, f_2, \ldots, f_q\}$, a set of q features (such as multimedia devices and experimental facilities) of the rooms.
- $Z = \{z_1, z_2, \ldots, z_w\}$, a set of w teachers.
- $c(e_i)$, the capacity required by events e_i $(1 \leq i \leq n)$.
- $c(r_j)$, the capacity of rooms r_j $(1 \leq j \leq m)$.
- $A_{m \times q}$, a room-feature matrix where $A_{jl} = 1$ denotes that room r_j has feature f_l and $A_{jl} = 0$ otherwise $(1 \leq j \leq m; 1 \leq l \leq q)$.
- $B_{n \times q}$, an event-feature matrix where $B_{jl} = 1$ denotes that event e_i requires feature f_l and $B_{jl} = 1$ otherwise $(1 \leq j \leq m; 1 \leq l \leq q)$.
- $C_{n \times n}$, an event-conflict matrix where $C_{ii'}$ denotes whether two events e_i and $e_{i'}$ can be held in the same timeslot $(1 \leq i \leq n; 1 \leq i' \leq n)$.
- $D_{p \times n}$, a student-event matrix where D_{hi} denotes whether students s_h will attend event e_i $(1 \leq h \leq p; 1 \leq i \leq n)$.
- $G_{w \times n}$, a teacher-event matrix where G_{vi} denotes whether teachers z_v will attend event e_i $(1 \leq v \leq w; 1 \leq i \leq n)$.

Commonly, we have $o = 12 \times 5$ timeslots, where 5 is the number of working days and 12 is the number of classes in each working day.

The decision variables of the problem is a three-dimensional event-room-timeslot matrix $X_{(n \times m \times o)}$, where X_{ijk} indicates that event e_i is to be held in room r_j and timeslot t_k $(1 \leq i \leq n; 1 \leq j \leq m; 1 \leq k \leq o)$.

For convenience, let D_h denote the set of events of students s_h, G_v be the set of events of teacher z_v, X_j be the set of events assigned to room r_j, C_i be the set of events conflicting to event e_i, and B_i be the set of features required by event e_i.

The problem has the following hard constraints.

- Each event must be scheduled once and only once:

$$\sum_{j=1}^{m} \sum_{k=1}^{o} X_{ijk} = 1, \quad 1 \leq i \leq n \tag{7.36}$$

- Each student can attend one event in any timeslot:

$$\sum_{i \in D_h} \sum_{j=1}^{m} X_{ijk} \leq 1, \quad 1 \leq h \leq p; 1 \leq k \leq o \tag{7.37}$$

- Each room can hold one event in any timeslot:

$$\sum_{i \in X_j} X_{ijk} \leq 1, \quad 1 \leq j \leq m; 1 \leq k \leq o \tag{7.38}$$

- Any two conflicting events cannot be held in the same timeslot:

$$\sum_{i' \in C_i} \sum_{j=1}^{m} X_{i'jk} \leq 1, \quad 1 \leq i \leq n; 1 \leq k \leq o \tag{7.39}$$

- Each room allocated to an event should have adequate features:

$$c(e_i) \leq c(r_j), \quad \forall i, j : (\exists k : X_{ijk} = 1) \tag{7.40}$$

- Each room allocated to an event should have adequate features:

$$A_{jl} = 1, \quad \forall i, j, l : (\exists k : X_{ijk} = 1) \wedge l \in B_i \tag{7.41}$$

Note that Eq. (7.37) is a hard constraint on the timeslots allocated to students, but there is no similar hard constraint on teachers, because such conflicts have already been specified in $C_{(n \times n)}$ and thus constrained by Eq. (7.39).

The problem has the following soft constraints.

- Each student should not take more than four timeslots in a day:

$$\sum_{i \in D_h} \sum_{j=1}^{m} \sum_{k=a}^{a+11} X_{ijk} \leq 4, \quad 1 \leq h \leq p; a \in \{1, 13, 25, 37, 49\} \tag{7.42}$$

- Each student should not take more than two consecutive timeslots in a day:

$$\sum_{i \in D_h} \sum_{j=1}^{m} \sum_{k=a}^{a+2} X_{ijk} \leq 2, \quad 1 \leq h \leq p; 1 \leq a \leq 60 \wedge a\%12 \leq 10 \tag{7.43}$$

- Each teacher should not teach more than four timeslots in a day:

$$\sum_{i \in G_v} \sum_{j=1}^{m} \sum_{k=a}^{a+11} X_{ijk} \leq 4, \quad 1 \leq v \leq w; a \in \{1, 13, 25, 37, 49\} \tag{7.44}$$

- Each teacher should not teach more than two consecutive timeslots in a day:

$$\sum_{i \in G_v} \sum_{j=1}^{m} \sum_{k=a}^{a+2} X_{ijk} \leq 2, \quad 1 \leq v \leq w; 1 \leq a \leq 60 \wedge a\%12 \leq 10 \tag{7.45}$$

- Each event should not be assigned to the final timeslot of each day:

$$\sum_{i=1}^{n} \sum_{j=1}^{m} X_{ija} = 0, \quad a \in \{12, 24, 36, 48, 60\} \tag{7.46}$$

The problem is to find a schedule of events such that the hard constraints (7.36)–(7.41) are all satisfied (otherwise, the solution is infeasible), while the soft constraints (7.42)–(7.46) are violated as minimum as possible (so as to improve the solution quality). Let $S_1 = \{1, 13, 25, 37, 49\}$, $S_2 = \{a|1 \leq a \leq 60 \wedge a\%12 \leq 10\}$, and $S_3 = \{12, 24, 36, 48, 60\}$, the objective function of the problem is defined as:

$$
\begin{aligned}
\min f = & \sum_{a \in S_1} \sum_{h=1}^{p} (\sum_{i \in D_h} \sum_{j=1}^{m} \sum_{k=a}^{a+11} g(X_{ijk} - 4)) \\
& + \sum_{a \in S_2} \sum_{h=1}^{p} (\sum_{i \in D_h} \sum_{j=1}^{m} \sum_{k=a}^{a+2} g(X_{ijk} - 2)) \\
& + \sum_{a \in S_1} \sum_{v=1}^{w} (\sum_{i \in G_v} \sum_{j=1}^{m} \sum_{k=a}^{a+11} g(X_{ijk} - 4)) \\
& + \sum_{a \in S_2} \sum_{v=1}^{w} (\sum_{i \in G_v} \sum_{j=1}^{m} \sum_{k=a}^{a+2} g(X_{ijk} - 2)) \\
& + \sum_{a \in S_3} (\sum_{i=1}^{n} \sum_{j=1}^{m} X_{ija})
\end{aligned}
\tag{7.47}
$$

where g is a function defined as

$$
g(x) = \begin{cases} x, & x \geq 0 \\ 0, & x < 0 \end{cases}
\tag{7.48}
$$

7.5.2   A Discrete EBO Algorithm for UCTP

From the above UCTP formulation, we can see that evaluating either the feasibility or the objective function of a solution is often computationally expensive. Thus, we propose an efficient discrete EBO algorithm for the complex problem [39].

Solution encoding and constraint handling Directly encoding a solution as an $n \times m \times o$ matrix can be very inefficient. Here, we employ a two-stage procedure for constructing and evaluating each UCTP solution as follows:

(1) Determining a timeslot for each event.
(2) Using the bipartite matching algorithm [15] to determine a room for each event with specified timeslot.

This enables us to use a compact encoding: Each solution $\mathbf{x}$ is a n-dimensional integer valued vector, where $x(i)$ denotes the timeslot allocated for event e_i. Such a vector is obtained in stage 1 and is given as an input to stage 2. Thereby, we can easily transform $\mathbf{x}$ to a matrix $X_{n \times m \times o}$ by using the procedure shown in Algorithm 7.8, where y denotes the result of stage 2 and $y(i)$ denotes the room assigned for event e_i with the timeslot determined in stage 1.

Algorithm 7.8: The decoding procedure of the discrete EBO for UCTP

---

1 Perform the bipartite matching algorithm on x to generate y;
2 Initialize a three-dimensional matrix $X_{n \times m \times o}$;
3 **for** $i = 1$ *to* n **do**
4   **for** $j = 1$ *to* m **do**
5   **for** $k = 1$ *to* l **do**
6   **If** $x(i) = k \wedge y(i) = j$ **then** $X_{ijk} = 1$;
7   **else** $X_{ijk} = 0$

8 **return** X;

Note that the compact encoding ensures the satisfaction of constraint (7.36), and the result of the bipartite matching algorithm also satisfies the constraints (7.38), (7.40) and (7.41). So we use the following penalized objective function that just considers the violations of constraints (7.37) and (7.39), where M is a large positive constant (typically larger than the maximum possible number of soft constraint violations):

$$f' = f + M \left(\sum_{h=1}^{p} \sum_{k=1}^{o} \left(\sum_{i \in D_h} \sum_{j=1}^{m} g(X_{ijk}) - 1 \right) + \sum_{i=1}^{n} \sum_{k=1}^{o} \left(\sum_{i' \in C_i} \sum_{j=1}^{m} g(X_{i'jk}) - 1 \right) \right)$$

(7.49)

As UCTP is a combinatorial problem, we need to redefine the local and global migration operators in the discrete EBO.

Local migration When performing local migration on solution $\mathbf{x}_i$, we select a neighboring solution $\mathbf{x}_j$ with a probability in proportional to μ_j and select a random dimension $k \in [1, n]$, and then set $\mathbf{x}_i(k)$ to $\mathbf{x}_j(k)$. Let $u = \mathbf{x}_j(k)$, we then perform the following procedure to repair $\mathbf{x}_i$:

(1) Sort the set of timeslots other than u in increasing order of the number of events assigned to each timeslot, and let $T(\mathbf{x}_i)$ be the sorted list.
(2) If $T(\mathbf{x}_i)$ is empty, stop the repair procedure.
(3) Otherwise, check whether there exists another position k' of $\mathbf{x}$ such that $\mathbf{x}_i(k') = u$; if so, perform the following steps:

    3.1 Check whether $C_{k,k'} = 1$; if so, go to Step 3.4.
    3.2 Otherwise, check whether there exists any student s_h such that $D_{h,k} = 1$ and $D_{h,k'} = 1$; if so, go to Step 3.4.
    3.3 Otherwise, check whether there exists any teacher z_v such that $G_{v,k} = 1$ and $G_{v,k'} = 1$; if so, go to Step 3.4, otherwise stop the repair procedure.
    3.4 Let u be the first timeslot of $T(\mathbf{x}_i)$, remove u from $T(\mathbf{x}_i)$, set $\mathbf{x}_i(k') = u$, let $k = k'$, and go to Step 2.

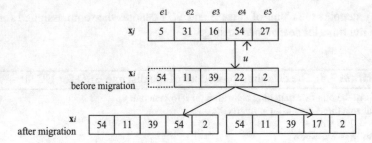

Fig. 7.6 An example of local migration for UCTP

In case that $\mathbf{x}_i$ is a feasible solution, if the above procedure stops at Step 3.3, the new solution is also feasible, and thus when evaluating its objective function, we do not need to calculate the penalized part of Eq. (7.49); otherwise, the new solution is an infeasible one and Eq. (7.49) must be completely evaluated. Thus, the procedure can greatly reduce the computational burden.

Figure 7.6 shows an example of the above local migration process. Assuming $k = 4$, we first set $u = \mathbf{x}_j(4) = 54$, as we can see in $\mathbf{x}_i(1) = u$, so we should check whether event e_1 and e_4 can be held in the same timeslot. If this migration will not lead to any constraint violation, we directly migrate u from $\mathbf{x}_j$ to $\mathbf{x}_i$, which denotes e_4 is reassigned to timeslot 54. Otherwise, we perform the repair process to set u to the timeslot that has been allocated with the minimum number of events. Let timeslot 17 to be the first element of $T(\mathbf{x}_i)$, after migration e_4 is reassigned to timeslot 17.

Global migration When performing global migration on solution $\mathbf{x}_i$, we select a neighbor $\mathbf{x}_j$ and a non-neighbor $\mathbf{x}_{j'}$ with probabilities in proportional to their emigration rates and select two different dimensions $k, k' \in [1, n]$, and then perform the following steps:

(1) Set $\mathbf{x}_i(k)$ to $\mathbf{x}_j(k)$, and then perform the above repair procedure at the kth dimension. If the result is infeasible, then go to Step 3.
(2) Set $\mathbf{x}_i(k')$ to $\mathbf{x}_{j'}(k')$, and then perform the repair procedure at the k'th dimension. If the result is feasible, then return the resulting solution.
(3) Set $\mathbf{x}_i(k)$ to $\mathbf{x}_{j'}(k)$, and then perform the repair procedure at the kth dimension. If the result is feasible, then return the resulting solution.
(4) Set $\mathbf{x}_i(k')$ to $\mathbf{x}_j(k')$, and then perform the repair procedure at the k'th dimension.

Figure 7.7 shows two examples of the global migration. The first example in Fig. 7.7a shows that the solution $\mathbf{x}_i$ accepts features from the neighbor $\mathbf{x}_j$ at the kth dimension and from the non-neighbor $\mathbf{x}_{j'}$ at the k'th dimension, which means that event e_2 is reassigned to timeslot 34 and event e_4 is reassigned to timeslot 54. The second example in Fig. 7.7b shows that $\mathbf{x}_i$ accepts features from $\mathbf{x}_j$ at the k'th dimension and from $\mathbf{x}_{j'}$ at the kth dimension, which means that e_2 is reassigned to timeslot 31 and event e_4 is reassigned to timeslot 14. Note that if the immigrating component is already in another dimension of the current solution, the solution still needs to be repaired.

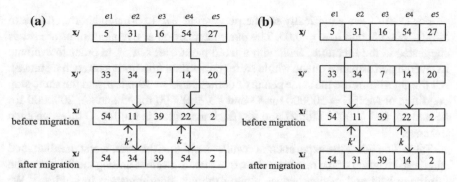

Fig. 7.7 Two examples of global migration for UCTP

Table 7.5 Summary of the test instances of UCTP [39]

No.	n	m	o	p	q	w
1	156	8	60	285	3	15
2	172	8	60	281	3	15
3	462	26	60	633	5	30
4	472	26	60	670	5	28
5	689	37	60	1475	6	53
6	706	37	60	1543	6	66
7	1711	72	60	5125	6	151
8	1860	72	60	5203	6	169

The discrete EBO algorithm uses the same migration model and algorithmic framework Algorithm 4.1 as the original EBO, except that the local and global migrations are modified as two mentioned above.

7.5.3 Computational Experiments

The discrete EBO (D-EBO) algorithm is tested on eight UCTP instances summarized in Table 7.5, which include four small- and medium-size instances (#1–#4) and four large-size instances (#5–#8). The following five algorithms are used for comparison:

- A GA algorithm [1].
- A ACO algorithm [29].
- A hybrid PSO algorithm [37].
- A variable neighborhood search (VNS) algorithm [7].
- A discrete BBO algorithm that only uses the local migration described in Sect. 7.5.2 (denoted by D-BBO).

For D-EBO, we empirically set the population size to 50, neighborhood size to 5, $\eta_{max} = 0.95$, and $\eta_{min} = 0.05$. The parameters of other five algorithms are set as suggested in the literature, each with a fixed parameter setting in order to evaluate its overall performance on the whole suite (rather than fine-tuning on each instance). All the algorithms use the same penalty coefficient $M = 10nmop$, and the same stop condition of MNFE $= 500000$ for #1 and #2, 1000000 for #3 and #4, 2000000 for #5 and #6, and 4000000 for #7 and #8. Each algorithm is run for 30 times on each instance.

Table 7.6 shows the experimental results, including the best, worst, median, and standard deviation of the resulting objective function values among the 30 runs. The minimum best and median values among the six algorithms are in boldface. We have also conducted paired t tests between D-EBO and each of the other algorithms, and mark $^+$ before the median values in columns 3–7 if D-EBO has statistically significant performance improvement over the corresponding algorithms (at 95% confidence level).

As we can see, on the first two small-size instances, only D-EBO and D-BBO can obtain solution(s) that satisfy all the hard and soft constraints in all 30 runs (where both the median and best objective values are zero). On instance #3, all the algorithms can obtain the best objective value of 6, but only D-EBO can obtain a median value of 6 which is much less than the other algorithms. On the remaining five instances, D-EBO not only always obtains the minimum median values but also uniquely obtains the minimum best values. Particularly, on the last four large-size instances, the results of D-EBO are much better than the other five algorithms. In terms of statistical tests, except that on #1–#3 there is no significant difference between D-EBO and D-BBO, D-EBO has statistically significant improvement over the other algorithms on all the test instances. This validates the effectiveness of the specific migration operator design for UCTP.

Among the other five algorithms, D-BBO achieves the minimum best and median values on six instances, which indicates that the local migration described in Sect. 7.5.2 is much more effective than operators such as the crossover and mutation of GA and the solution path construction method of ACO on the test problems. Nevertheless, D-EBO exhibits a significant performance improvement over D-BBO, which shows that the combination of local and global migration is further more effective than the single local migration operator for UCTP. In particular, the VNS algorithm used in our experiments is an enhanced version of the third ITC winner algorithm. Although its results are better than D-BBO on two instances (#6 and #7), the results of D-BBO are comparable to those of VNS on the other instances. More importantly, VNS never performs better than D-EBO on any instance, mainly because VNS focuses on local search, whereas D-EBO utilizes the neighborhood structure in collaboration with migration to enhance the interactions between individuals in the population and thus enable the algorithm to achieve a better balance between exploration and exploitation.

Figure 7.8 presents the convergence curves of the algorithms on the UCTP instances, where the x-axis denotes the consumed NFEs (in 10^3) and the y-axis denotes the median objective values. It can be observed that both the convergence speeds and the final results of D-EBO are much better than the other five algorithms

Table 7.6   Comparative results on the test UCTP instances [39]

#	Metric	GA	ACO	PSO	VNS	D-BBO	D-EBO
1	Best	**0**	**0**	**0**	**0**	**0**	**0**
	Worst	12	12	12	5	0	0
	Median	5.67^+	5.33^+	5.93^+	1.67^+	**0**	**0**
	Std	3.13	2.48	3.02	2.11	0	0
2	Best	**0**	**0**	**0**	**0**	**0**	**0**
	Worst	16	12	16	12	9	0
	Median	6.87^+	5.13^+	7.00^+	4.97^+	**0**	**0**
	Std	4.05	3.93	3.95	3.65	3.27	0
3	Best	**6**	**6**	**6**	**6**	**6**	**6**
	Worst	21	19	26	21	19	6
	Median	15.50^+	15.00^+	15.13^+	10.50^+	6.37	**6.00**
	Std	3.97	3.66	5.2	3.88	2.75	0
4	Best	25	25	25	25	21	**14**
	Worst	39	39	39	39	39	21
	Median	33.03^+	31.67^+	26.90^+	27.53^+	24.50^+	**16.03**
	Std	4.12	3.89	4.48	3.70	3.90	3.18
5	Best	73	68	72	65	65	**53**
	Worst	105	105	123	105	105	79
	Median	93.00^+	89.33^+	98.17^+	90.20^+	82.77^+	**58.30**
	Std	16.79	16.50	19.19	17.80	15.73	5.88
6	Best	170	166	132	93	109	**72**
	Worst	261	248	259	252	230	133
	Median	220.53^+	202.20^+	180.00^+	150.63^+	161.37	**98.13**
	Std	30.35	37.92	39.15	38.03	42.16	25.32
7	Best	305	290	279	233	271	**185**
	Worst	494	468	510	387	411	229
	Median	375.83^+	338.23^+	360.50^+	316.00^+	330.93	**216.60**
	Std	50.08	48.17	83.29	57.25	57.9	28.71
8	Best	453	435	385	376	360	**199**
	Worst	610	626	579	515	521	360
	Median	538.10^+	589.50^+	436.27^+	424.60^+	418.13^+	**255.87**
	Std	55.20	50.55	63.99	52.17	57.9	37.68

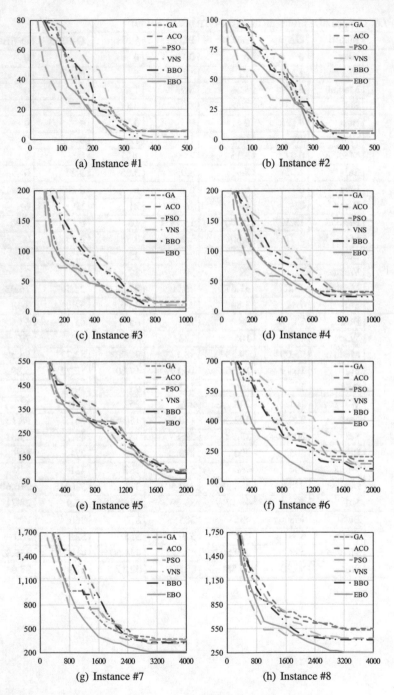

Fig. 7.8 Convergence curves of the six algorithms on the UCTP instances [39]

on almost all the instances. Although in early search stages PSO converges faster than D-EBO, it soon slows down and is overtaken by D-EBO at later stages, because PSO particles converge quickly to a small area by learning from the global best and lose the chance of jumping out of the local optimum if the global best is a local optimum, while in D-EBO habitats still have the opportunity to perform the global migration in the later stages to explore wider areas of the search space and find better solutions. Compared with D-BBO, on most instances D-EBO converges faster during the whole search process and finally reaches better results, which validates that EBO can outperform the basic BBO not only on continuous optimization problems but also on the discrete UCTP.

7.6 Summary

This chapter introduces BBO and its improved versions for a set of job scheduling problems. They involve two strategies for tackling with combinatorial optimization problems: For FSP, we use a real-value solution encoding with the original BBO operators by applying specific encoding/decoding between the real vectors and discrete solutions; For the other problems, we design special migration and mutation operators for the discrete solution space.

References

1. Alsmadi OMK, Abo-Hammour ZS, Abu-Al-Nadi DI, Algsoon A (2011) A novel genetic algorithm technique for solving university course timetabling problems. In: International workshop on systems, signal processing and their applications, pp 195–198. https://doi.org/10.1109/WOSSPA.2011.5931449
2. Banharnsakun A, Sirinaovakul B, Achalakul T (2012) Job shop scheduling with the best-so-far abc. Eng Appl Artif Intel 25(3):583–593. https://doi.org/10.1016/j.engappai.2011.08.003
3. Carter AE, Ragsdale CT (2006) A new approach to solving the multiple traveling salesperson problem using genetic algorithms. Eur J Oper Res 175:246–257. https://doi.org/10.1016/j.ejor.2005.04.027
4. Chang YL, Matsuo H, Sullivan R (1989) A bottleneck-based beam search for job scheduling in a flexible manufacturing system. Int J Prod Res 27:1949–1961. https://doi.org/10.1080/00207548908942666
5. Deb K, Pratap A, Agarwal S, Meyarivan T (2002) A fast and elitist multiobjective genetic algorithm: NSGA-II. IEEE Trans Evol Comput 6:182–197. https://doi.org/10.1109/4235.996017
6. Didier Dubois HP (1978) Operations on fuzzy numbers. Int J Syst Sci, pp 613–626. https://doi.org/10.1080/00207727808941724
7. Fonseca GHG, Santos HG (2014) Variable neighborhood search based algorithms for high school timetabling. Comput Oper Res 52:203–208. https://doi.org/10.1016/j.cor.2013.11.012
8. Framinan J, Gupta J, Leisten R (2004) A review and classification of heuristics for permutation flow-shop scheduling with makespan objective. J Oper Res Soc 55:1243–1255. https://doi.org/10.1057/palgrave.jors.2601784
9. Garey M, Johnson D (1979) Computers and intractability. A guide to the theory of NP-completeness. https://doi.org/10.2307/2273574

10. Garey MR, Johnson DS, Sethi R (1976) The complexity of flowshop and jobshop scheduling. Math Oper Res 1:117–129. https://doi.org/10.1287/moor.1.2.117
11. Ge HW, Sun L, Liang YC, Qian F (2008) An effective pso and ais-based hybrid intelligent algorithm for job-shop scheduling. IEEE Trans Syst Man Cybern B. Cybern 38:358–368. https://doi.org/10.1109/TSMCA.2007.914753
12. Goncalves J, Mendes J, Resende M (2005) A hybrid genetic algorithm for the job shop scheduling problem. Eur J Oper Res 167:77–95. https://doi.org/10.1016/j.ejor.2004.03.012
13. Gupta J, Stafford E (2006) Flowshop scheduling research after five decades. Eur J Oper Res 169:699–711. https://doi.org/10.1016/j.ejor.2005.02.001
14. Hajek J, Szollos A, Sistek J (2010) A new mechanism for maintaining diversity of pareto archive in multi-objective optimization. Adv Eng Softw 41:1031–1057. https://doi.org/10.1016/j.advengsoft.2010.03.003
15. Hopcroft JE (1973) An n5/2 algorithm for maximum matchings in bipartite graphs. SIAM J Comput 2:122–125. https://doi.org/10.1137/0202019
16. Hu J, Mehrotra S (2012) Robust and stochastically weighted multiobjective optimization models and reformulations. Oper Res 60:936–953. https://doi.org/10.1287/opre.1120.1071
17. Knowles J, Watson R, Corne D (2001) Reducing local optima in single-objective problems by multi-objectivization. In: Evolutionary multi-criterion optimization, Lecture Notes in Computer Science, vol 1993, pp 269–283. https://doi.org/10.1007/3-540-44719-9-19
18. Krasnogor N, Smith J (2000) A memetic algorithm with self-adaptive local search: TSP as a case study. In: Proceedings of 2nd Annual Conference on Genetic and Evolutionary Computation, pp 987–994
19. Lawler EL, Lenstra JK, Rinnooy Kan AHG (1982) Recent developments in deterministic sequencing and scheduling: A survey. In: Deterministic and stochastic scheduling, pp 35–73. https://doi.org/10.1007/978-94-009-7801-0-3
20. Lawrence S (1984) Supplement to resource constrained project scheduling: an experimental investigation of heuristic scheduling techniques. Energy Procedia 4(7):4411–4417
21. Lin TL, Horng SJ, Kao TW, Chen YH, Run RS, Chen RJ, Lai JL, Kuo IH (2010) An efficient job-shop scheduling algorithm based on particle swarm optimization. Expert Syst Appl 37:2629–2636. https://doi.org/10.1016/j.eswa.2009.08.015
22. Liu B, Liu Y (2002) Expected value of fuzzy variable and fuzzy expected value models. IEEE Trans Fuzzy Syst 10:445–450. https://doi.org/10.1109/TFUZZ.2002.800692
23. Liu B, Wang L, Jin YH (2007) An effective pso-based memetic algorithm for flow shop scheduling. IEEE Trans Syst Man Cybern B. Cybern 37(1):18–27. https://doi.org/10.1109/TSMCB.2006.883272
24. Ma H, Simon D (2011) Blended biogeography-based optimization for constrained optimization. Eng Appl Artif Intel 24:517–525. https://doi.org/10.1016/j.engappai.2010.08.005
25. Matsatsinis N, Samaras A (2001) Mcda and preference disaggregation in group decision support systems. Eur J Oper Res 130:414–429. https://doi.org/10.1016/S0377-2217(00)00038-2
26. Michael P (1995) Scheduling: theory, algorithms, and systems
27. Moscato P, Cotta C (2003) A gentle introduction to memetic algorithms. In: Handbook of Metaheuristics, pp 105–144. https://doi.org/10.1007/0-306-48056-5_5
28. Nawaz M, Enscore EE Jr, Ham I (1983) A heuristic algorithm for the m-machine, n-job flowshop sequencing problem. Omega 11:91–95. https://doi.org/10.1016/0305-0483(83)90088-9
29. Nothegger C, Mayer A, Chwatal A, Raidl GR (2012) Solving the post enrolment course timetabling problem by ant colony optimization. Ann Oper Res 194:325–339. https://doi.org/10.1007/s10479-012-1078-5
30. Nowicki E, Smutnicki C (1996) A fast taboo search algorithm for the job shop problem. Manag Sci 42:797–813. https://doi.org/10.1287/mnsc.42.6.797
31. Qian B, Wang L, Hu R, Wang WL, Huang DX, Wang X (2008) A hybrid differential evolution method for permutation flow-shop scheduling. Int J Adv Manuf Technol 38(7–8):757–777. https://doi.org/10.1007/s00170-007-1115-8
32. Reeves CR, Yamada T (1998) Genetic algorithms, path relinking, and the flowshop sequencing problem. Evol Comput 6(1):45–60. https://doi.org/10.1162/evco.1998.6.1.45

33. Simon D (2008) Biogeography-based optimization. IEEE Trans Evol Comput 12:702–713. https://doi.org/10.1109/TEVC.2008.919004
34. Sniedovich M (2012) Black swans, new nostradamuses, voodoo decision theories, and the science of decision making in the face of severe uncertainty. Int Trans Oper Res 19:253–281. https://doi.org/10.1111/j.1475-3995.2010.00790.x
35. Spanos AC, Ponis ST, Tatsiopoulos IP, Christou IT, Rokou E (2014) A new hybrid parallel genetic algorithm for the job-shop scheduling problem. Int Trans Oper Res 21:479–499. https://doi.org/10.1111/itor.12056
36. Storer RH, Wu SD, Vaccari R (1992) New search spaces for sequencing problems with application to job shop scheduling. Manag Sci 38(10):1495–1509. https://doi.org/10.1287/mnsc.38.10.1495
37. Tassopoulos IX, Beligiannis GN (2012) A hybrid particle swarm optimization based algorithm for high school timetabling problems. Appl Soft Comput 12:3472–3489. https://doi.org/10.1016/j.asoc.2012.05.029
38. Yin M, Li X (2011) A hybrid bio-geography based optimization for permutation flow shop scheduling. Sci Res Essays 6. https://doi.org/10.5897/SRE10.818
39. Zhang MX, Zhang B, Qian N (2017) University course timetabling using a new ecogeography-based optimization algorithm. Nat Comput 16(1):61–74. https://doi.org/10.1007/s11047-016-9543-8
40. Zheng Y, Shi H, Chen S (2012) Fuzzy combinatorial optimization with multiple ranking criteria: a staged tabu search framework. Pac J Optim 8:457–472
41. Zheng Y, Chen S, Ling H (2013) Efficient multi-objective tabu search for emergency equipment maintenance scheduling in disaster rescue. Opt Lett 7:89–100. https://doi.org/10.1007/s11590-011-0397-9
42. Zheng YJ, Ling HF, Wu XB, Xue JY (2014) Localized biogeography-based optimization. Soft Comput 18:2323–2334. https://doi.org/10.1007/s00500-013-1209-1
43. Zheng YJ, Ling HF, Xue JY (2014) Ecogeography-based optimization: enhancing biogeography-based optimization with ecogeographic barriers and differentiations. Comput Oper Res 50:115–127. https://doi.org/10.1016/j.cor.2014.04.013
44. Zheng YJ, Ling HF, Xu XL, Chen SY (2015) Emergency scheduling of engineering rescue tasks in disaster relief operations and its application in china. Int Trans Oper Res 22:503–518. https://doi.org/10.1111/itor.12148

Chapter 8
Application of Biogeography-Based Optimization in Image Processing

Abstract Computer image processing gives rise to many optimization problems, which can be very difficult when the images are large and complex. In this chapter, we use BBO and its improved versions to a set of optimization problems in image processing, including image compression, salient object detection, and image segmentation. The results demonstrate the effectiveness of BBO in optimization problems in image processing.

8.1 Introduction

There are a variety of optimization problems in the field of image processing. This chapter introduces the applications of BBO in image processing, including image compression, salient object detection, and image segmentation. The results demonstrate the effectiveness of BBO in image processing optimization.

8.2 BBO for Image Compression

8.2.1 Fractal Image Compression

The digital representation of an image often requires a large amount of data, which can be inconvenient to store and transmit. Image compression aims to reduce the data amount without or with a minimum loss of image information. In general, image compression is based on two facts:

- There are often a lot of duplicate data in image information, which could be represented much more concisely by some mathematical techniques.

- There are a lot of image details that cannot be recognized by human eyes, and the removal of such details would have little influence on the image quality felt by humans.

Image compression based on the first fact is called "lossless compression," which allows us to completely recover the original data from the compressed one. For example, the TIFF image file format uses lossless compression. By contrast, image compression based on the second fact is called "lossy compression," where we can only recover parts of the original data. For example, the JPEG image file format uses lossy compression.

Fractal image compression (FIC) is a lossy compression method, which utilizes the similarity between different parts of an image. The advantages of FIC include high compression ratio and fast decoding speed, but its main disadvantage is the high computational cost used to find similar parts. FIC is theoretically based on the following three theorems [6]:

- The iterated function system (IFS) theorem indicating that each IFS forms a contraction mapping in the function space.
- The fixed-point theorem indicating that each contraction mapping has a fixed point.
- The collage theorem indicating that an image can be transformed into L image sets through L contraction mappings. The L small image sets can be collaged into a new image; if the difference between the new image and the original image is arbitrarily small, the IFS constructed is arbitrarily close to the original image.

IFS uses affine transformation [11, 17]. In a 2D space, an affine transformation is equivalent to a linear transformation (of position vectors) followed by a translation. In a 2D grayscale image, each pixel point is represented by three values (x, y, z), where z is grayscale value in addition to the 2D location (x, y), and thus, an affine transformation ω can be defined as:

$$\omega \begin{bmatrix} x \\ y \\ z \end{bmatrix} = \begin{bmatrix} a_{11} & a_{12} & 0 \\ a_{21} & a_{22} & 0 \\ 0 & 0 & s \end{bmatrix} \begin{bmatrix} x \\ y \\ z \end{bmatrix} + \begin{bmatrix} \Delta x \\ \Delta y \\ o \end{bmatrix} \tag{8.1}$$

where s is a scale factor, o is an offset factor, $\mathbf{B} = [\Delta x, \Delta y]^{\mathrm{T}}$ is an offset vector, and $\mathbf{A} = \begin{bmatrix} a_{11} & a_{12} \\ a_{21} & a_{22} \end{bmatrix}$ is a transformation matrix taking one of the following forms:

$$\begin{bmatrix} 1 & 0 \\ 0 & 1 \end{bmatrix}, \begin{bmatrix} 1 & 0 \\ 0 & -1 \end{bmatrix}, \begin{bmatrix} -1 & 0 \\ 0 & 1 \end{bmatrix}, \begin{bmatrix} -1 & 0 \\ 0 & -1 \end{bmatrix}, \begin{bmatrix} 0 & 1 \\ 1 & 0 \end{bmatrix}, \begin{bmatrix} 0 & -1 \\ 1 & 0 \end{bmatrix}, \begin{bmatrix} 0 & 1 \\ -1 & 0 \end{bmatrix}, \begin{bmatrix} 0 & -1 \\ -1 & 0 \end{bmatrix} \tag{8.2}$$

The encoding process of FIC consists of the following four steps.

(1) Partitioning the original $(N \times N)$-dimensional image I is partitioned into $L \times L$ non-intersected blocks, where each block R_i is an encoding of in FIC.
(2) Creating a search space by using a $2L \times 2L$ window moves along the horizontal and vertical direction of the image L with the Δh and Δv steps, respectively.

Each captured area D_i is called as a domain block or matching block, and all these blocks compose the search space S_D. The size of the search S_D is:

$$|S_D| = \left(\frac{N - 2L}{\Delta h} + 1\right) \times \left(\frac{N - 2L}{\Delta v} + 1\right) \tag{8.3}$$

(3) In the search space, searching the best matching block D_i for each range R_i, such that through affine transformation ω_i, D_i approximates R_i as much as possible:

$$d(R_i, \omega_i(D_i)) = \min_{D_i \in S_D} \|R_i - \omega_i(D_i)\|^2 \tag{8.4}$$

(4) Converting each ω_i into a fractal code, which is the compression result of R_i.

There are three typical ways to improve the performance of FIC: improving encoding speed, improving the quality of the decoded image (reduce information loss), and increasing the compression ratio [12]. From Eq. (8.3), the time complexity of searching a matching block is $O(N^2)$, so the time complexity of encoding the whole image is $O(N^4)$. For a larger image, this can be expensive, so here we consider improving the encoding speed.

In FIC, finding the best matching block is an optimization problem given by Eq. (8.4), and the problem can be divided into eight subproblems, each corresponding to one of the eight transformation matrices shown in Eq. (8.2). To minimize the mean square error in Eq. (8.4), the scale factor s_i and the offset factor o_i can be determined by the least square method:

$$s_i = \frac{\left(\sum_{x=0}^{L-1}\sum_{y=0}^{L-1} R_i(x, y) \cdot D_i(x, y)\right) - L^2 \overline{R}\,\overline{D}}{\left(\sum_{x=0}^{L-1}\sum_{y=0}^{L-1} D_i(x, y)\right) - L^2 \overline{D}} \tag{8.5}$$

$$o_i = \overline{R} - s_i \overline{D} \tag{8.6}$$

where

$$\overline{R} = \frac{1}{L^2}\sum_{x=0}^{L-1}\sum_{y=0}^{L-1} R_i(x, y) \tag{8.7}$$

$$\overline{D} = \frac{1}{L^2}\sum_{x=0}^{L-1}\sum_{y=0}^{L-1} D_i(x, y) \tag{8.8}$$

Thus, the decision variables are the offset vector $\mathbf{B} = [\Delta x, \Delta y]^T$. Take a 256×256 image and an 8×8 block as an example, each component value is in the range of

[0, 240], and the problem of finding the best matching block in FIC can be regarded as an integer programming problem.

8.2.2   BBO Algorithms for Fractal Image Compression

When applying BBO to solve the above problem, each solution (habitat) represents a two-dimensional vector $(\Delta x, \Delta y)$. The basic framework of BBO for FIC is shown in Algorithm 8.1. The algorithm is iteratively applied to each range block until all the best matching blocks are found. We, respectively, use three versions of BBO, including the original BBO [20], B-BBO [16], and EBO [30] to implement the above framework for FIC. In B-BBO and EBO, any non-integer components is rounded to the nearest integer [24].

Algorithm 8.1: The BBO algorithm for fractal image compression

---

1 Randomly initialize a population of solutions;
2 **while** *the stop criterion is not satisfied* **do**
3   **for** *each solution $H = (\Delta x, \Delta y)$ in the population* **do**
4   Set the fitness $f_H = 0$;
5   **for** *each of the eight matrix in Eq.* (8.2) **do**
6   Calculate the scale factor s and the offset factor o;
7   Construct the affine transformation ω based on s and o;
8   Calculate the mapping $\omega(D)$ and its deviation d from the corresponding range block R;
9   **if** $1/d > f_H$ **then** $f_H \leftarrow 1/d$;
10   Calculate its migration and mutation rates of the solutions;
11   Update the best solution found H^*;
12   Perform migration and mutation on the solutions;
13 **return** H^*;

8.2.3   Computational Experiments

We use six images denoted by #1–#6, including two with size of 256×256, two of 512×512, and two of 1024×1024, and test the three BBO algorithms with the following algorithms:

- An FIC full search algorithm which is an exact algorithm [6].
- A GA for FIC [26].
- A PSO for FIC [23].

The population size is set to 35 for the PSO and 50 for the remaining algorithms. All other parameters are set as suggested in the literature. The five algorithms (except the full search) use the same stop condition that the NFE reaches the upper limit, which is set to 1000, 2000, and 3000 for 256×256, 512×512, and 1024×1024 images, respectively. Each algorithm is run for 30 times on each image, and the results are averaged over 30 runs and measured by the following metrics:

- T, the coding time(in seconds) of each function.
- SR, the speedup ratio of the heuristic algorithm to the FIC full search algorithm, i.e., the ratio of the running time of FIC full search to that of the heuristic algorithm.
- PSNR, the peak signal-to-noise ratio of the image:

$$PSNR = 10 \lg \frac{N-1}{MSE} \qquad (8.9)$$

where MSE denotes the mean square error of the image before and after compression. Obviously, a smaller MSE or a larger PSNR indicates less information loss of the image.

The experimental results are shown in Table 8.1, where the best SR and PSNR among the five heuristic algorithms are in boldface. As can be seen from the results, the SR of the heuristic algorithms is about $40 \sim 80$, which shows their significant speed advantage over the exact FIC full search algorithms. Among five heuristic algorithms, GA and BBO show poor performance. The SR of the PSO is always the largest due to its high computational efficiency. The SR of EBO is better than that of GA and BBO, but poorer than PSO and B-BBO, which is mainly due to the extra computational cost for maintaining the neighborhood structure. However, the compression quality of the EBO is the best (and its information loss is minimum). B-BBO shows a better balance between SR and PSNR. Roughly speaking, there are no statistically significant differences in running time between PSO, B-BBO, and EBO, and thus, EBO with the best PSNR is more preferred for FIC.

8.3 BBO for Salient Object Detection

8.3.1 Salient Object Detection

Salient object detection (SOD) is to detect the location of the most "salient" foreground object that captures most human perceptual attention in a scene [5]. It plays a crucial role in a broad range of computer vision and multimedia applications. For example, in intelligent transportation systems, SOD can be used to identify traffic signs, vehicles, and pedestrians [10]. In medical image processing, SOD facilitates the differentiation of pathological structures from neighboring anatomic structures [19]. In surveillance systems, SOD enables to detect unusual events such as traffic violations and unauthorized entry [22].

Table 8.1 Experimental results of FIC

No.	Measure	FIC	GA	PSO	BBO	B-BBO	EBO
#1	T	1988.5	50.5	47.1	48.5	48.7	49.7
	SR	1	39.4	**42.2**	41.0	40.8	40.0
	PSNR	28.7	26.8	27.9	27.5	28.0	**28.2**
#2	T	2100.5	52.0	47.7	49.0	49.2	50.3
	SR	1	40.4	**44.0**	42.9	42.7	41.6
	PSNR	28.5	26.7	27.7	27.2	27.8	**28.2**
#3	T	7533.3	146.7	113.9	126.8	117.1	118.8
	SR	1	51.4	**66.1**	59.4	64.3	63.4
	PSNR	28.2	26.2	27.3	26.9	27.6	**28.0**
#4	T	7720.8	150.06	115.3	128.9	117.9	119.6
	SR	1	51.5	**67.0**	59.9	65.5	64.6
	PSNR	28.1	26.2	27.3	26.8	27.6	**27.9**
#5	T	27853.6	428.5	339.3	392.1	347.6	360.0
	SR	1	65.0	**82.1**	71.0	80.1	77.4
	PSNR	27.3	25.9	27.0	26.5	27.3	**27.6**
#6	T	29109.9	445.6	343.3	396.7	351.9	363.1
	SR	1	65.3	**84.8**	73.4	82.7	80.2
	PSNR	26.9	25.8	26.9	26.4	27.1	**27.6**

The existing SOD methods can be classified into two categories: bottom-up methods and top-down methods [27]. In bottom-up methods, the differences between pixels or pitches on the low-level features such as color and edge are calculated, and saliency maps are extracted by assigning pixels that are more "different" from their surrounding areas with higher saliency values. Bottom-up methods are intuitive and fast, but they only work well for low-level saliency [5, 10]. In top-down methods, high-level knowledge about the objects to be detected is used to process saliency information in a task-dependent manner. Top-down methods have higher detection accuracies, but they require prior knowledge and their speed is low.

In [18], an approach that combines bottom-up and top-down methods is proposed. The approach consists of the following steps:

(1) Extract three feature maps from the original image based on multi-scale contrast, center-surround histogram, and color spatial distribution, denoted by f_1, f_2, and f_3, respectively.
(2) Determine a weight vector $\mathbf{w} = (w_1, w_2, w_3)$ satisfying $w_1 + w_2 + w_3 = 1$ for combining the three feature maps into tag a salient map S, where the saliency value of every pixel p in the pixel set P is calculated as:

$$S(p, w) = \sum_{k=1}^{3} w_k f_k(p) \tag{8.10}$$

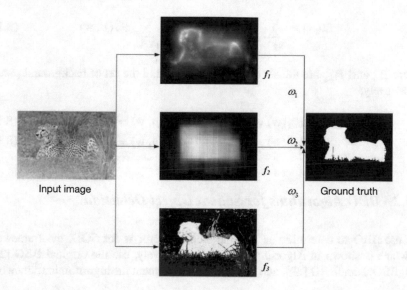

Fig. 8.1 A sample input image, showing maps f_1, f_2, and f_3 and its ground truth image [25]

(3) Normalize the saliency map S such that any the saliency value is in [0,1] (where 0 represents black and 1 represents white):

$$S_N(p, w) = \frac{S(p, \mathbf{w}) \quad \min_{p' \in P} S(p', \mathbf{w})}{\max_{p' \in P} S(p', \mathbf{w}) - \min_{p' \in P} S(p', \mathbf{w})} \tag{8.11}$$

where $\min_{p' \in P} S(p', \mathbf{w})$ and $\max_{p' \in P} S(p', \mathbf{w})$ denote the minimum and maximum saliency values in P.

(4) Use a threshold $\tau(\mathbf{w})$ to classify the salient pixels and background pixels, and thus, get a binarized attention mask A of salient objects:

$$A(p, w) = \begin{cases} 1, & S_N(p, \mathbf{w}) \geq \tau(\mathbf{w}) \\ 0, & S_N(p, \mathbf{w}) < \tau(\mathbf{w}) \end{cases} \tag{8.12}$$

where 1 denotes an attention pixel and 0 denotes a background pixel, and the threshold $\tau(\mathbf{w})$ is set to the half of the maximum value of the normalized saliency map S_N:

$$\tau(\mathbf{w}) = \frac{\max_{p' \in P} S(p', \mathbf{w})}{2} \tag{8.13}$$

Figure 8.1 illustrates the above process for SOD.

A key problem is to determine an appropriate weight vector $\mathbf{w}$ which differentiates attention pixels from background pixels as much as possible. This can be defined by the following fitness function of $\mathbf{w}$:

$$\text{fit}(\mathbf{w}) = \sum_{p \in P_{\text{att}}} (1 - S_N(p, \mathbf{w})) + \sum_{p \in P_{\text{bkg}}} S_N(p, \mathbf{w}) \tag{8.14}$$

where P_{att} and P_{bkg} are the set of attention pixels and the set of background pixels, respectively:

$$P_{\text{att}}(\mathbf{w}) = \{p : p \in P \wedge A(p, \mathbf{w}) = 1\} \tag{8.15}$$

$$P_{\text{bkg}}(\mathbf{w}) = \{p : p \in P \wedge A(p, \mathbf{w}) = 1\} \tag{8.16}$$

8.3.2   BBO Algorithms for Salient Object Detection

We use BBO to determine an optimal weight vector $\mathbf{w}$ for SOD, the framework of which is shown in Algorithm 8.2. We, respectively, use the original BBO [20], RingBBO, RandBBO [29], and EBO [30] to implement the algorithmic framework.

Algorithm 8.2: The BBO algorithm in the salient object detection

---

1 Randomly initialize a population of n solutions, each of which is a weight vector
   $\mathbf{w} = (w_1, w_2, w_3)$;
2 **while** *the stop criterion is not satisfied* **do**
3   **foreach w** *in the population* **do**
4   Build a saliency map S according to Eq. (8.10);
5   Normalize S to S_N according to Eq. (8.11);
6   Set the threshold value according to Eq. (8.13) and obtain the overall saliency map A
        according to Eq. (8.12);
7   Evaluate the fitness of $\mathbf{w}$ according to Eq. (8.14);
8   Calculate the migration and mutation rate of all solutions;
9   Update the best known solution found H^*;
10   Perform the migration mutation on the solutions;
11 **return** H^*;

8.3.3   Computational Experiments

We select 300 images from three widely used image data sets including MSRA [14], iCoSeg [3], and SED [1], which differ in attributes such as category, color, shape, and size. On these images besides, we compare our four BBO algorithms with the two other SOD algorithms, including the basic SOD model of Liu et al. [14] and a PSO algorithm by Singh et al. [21]. All BBO algorithms use the same population size of 20, and the parameters of the others are set as suggested in their publications.

We first run each BBO algorithm on each test image for five times. Base on the outputs of these algorithms, we divide the test images into three groups:

- Group A: Test images where all four BBO algorithms outputs are optimal.
- Group B: Test images where none of the four BBO algorithm outputs is optimal;
- Group C: Remaining test images, i.e., those there is some probability for BBO algorithms to achieve optimal salient results.

The values for the first group are averaged over five computational runs. For the other two groups, we carry out 20 additional simulations on each image and compute their average values for the top 20 of the total 25 runs.

The qualitative visual comparison of the best results for the six different algorithms on images in Groups A–C is shown in Figs. 8.2, 8.3, and 8.4, where the ground truth is marked on the input images. For images in Group B, Table 8.2 presents the average probability of achieving optimal salient results of each algorithm, which shows that the BBO algorithms have better results than the others, and the original BBO has the highest probability.

Clearly, Liu's method localizes the objects accurately, but it gives unnecessary information on the salient object, which reduces its performance. Singh's method obtains saliency results with precise profile information at times, but its probability to achieve optimal salient results is much smaller than BBO. With the increase of image complexity (e.g., images in the second row of Fig. 8.3 have only small texture differences between salient object and background), Singh's method is more likely to be trapped in local optima. BBO algorithms perform better with the complex images where the background is cluttered. For example, the images in the third row of Fig. 8.2 and images in the first two rows of Fig. 8.3 show that the other two approaches are distracted by the textures in the background, while BBO algorithms successfully output an accurate salient object. In Group C, BBO algorithms also produce very good results, confirming that they are more robust to the changes in color, object size, and object location in images with different background types.

Moreover, we use precision, recall, and the F-measure defined as follows to quantitatively evaluate the performance of the algorithms:

$$Precision = TP/(TP + FP) \tag{8.17}$$

$$Recall = TP/(TP + FN) \tag{8.18}$$

$$F\text{-measure} = \frac{2 \times Precision \times Recall}{Precision + Recall} \tag{8.19}$$

where TP denotes the true positives (the total number of attention pixels that are detected as salient), FP denotes the false positives (the number of background pixels that are detected as salient pixels), and FN denotes the false negatives (the number of attention pixels that are detected as background).

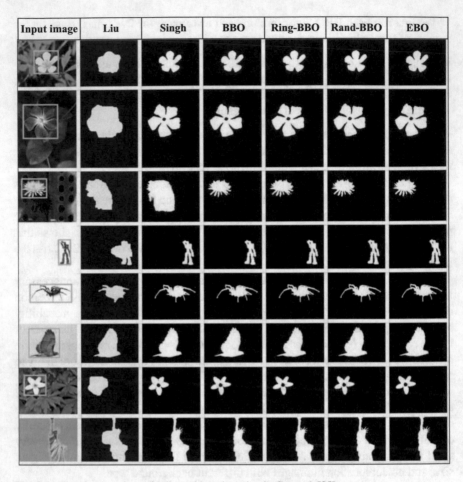

Fig. 8.2 Visual comparison of salient object detection in Group A [25]

Tables 8.3, 8.4, and 8.5, respectively, present the quantitative results of the above metrics as well as the computation time (in seconds) of the algorithms on Group A–C. It can be seen that all BBO algorithms have less computation time than Singh's and Liu's methods. Singh's method gives the highest recall rates, because it contains unnecessary information on regions surrounding the salient objects. On images in Group A, all the BBO algorithms achieve the highest precision and F-measure values. On images in Group B, RandBBO and EBO achieve the highest precision, and the EBO has the highest F-measure. On images in Group C, RandBBO has the highest precision and F-measure. In summary, the experimental results show that the BBO algorithms have a significant performance advantage over the others, and RandBBO exhibits the best performance among all the algorithms.

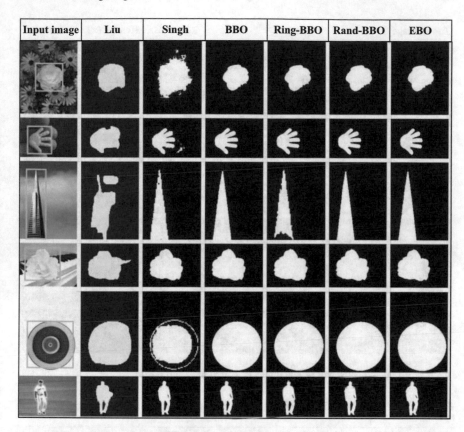

Fig. 8.3 Visual comparison of salient object detection in Group B [25]

8.4   BBO for Image Segmentation

8.4.1   Image Segmentation

Image segmentation is a process that partitions a raw image into a set of non-overlapping regions, each of which has different characteristics so that some interested objects can be obtained. The essence of image segmentation can be regarded as classifying/restructuring image pixels according to their characteristics. So it is not surprising that clustering methods have been widely used in image segmentation. K-means algorithm is one of the most classical clustering algorithms, the basic idea of which is dividing data points into k clusters and then continuously adjusting the points among the clusters to improve clusters with respect to the following objective function:

$$J_1 = \sum_{i=1}^{c} \sum_{k=1}^{n} (u_{ik}) \|x_k - v_i\| \tag{8.20}$$

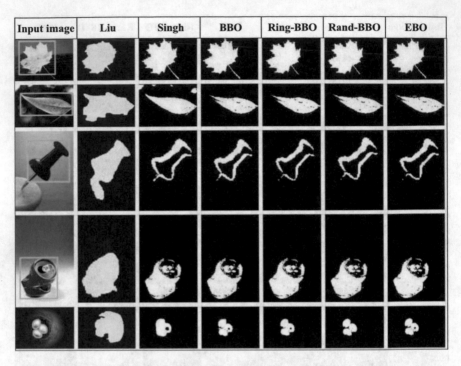

Fig. 8.4 Visual comparison of salient object detection in Group C [25]

Table 8.2 Probability of achieving optimal salient results in Group *B* images for various algorithms [25]

Algorithm	Liu	Singh	BBO	RingBBO	RandBBO	EBO
Probability (%)	0	23	**61**	47	56	54

where $u_{ik} = 1$ denotes that the kth sample point belongs to the ith class and $u_{ik} = 0$ otherwise. The K-means method is fast and easy to implement, but it is difficult to find appropriate clusters when the distribution shape is complex.

The fuzzy C-means (FCM) clustering [4] is an improvement of K-means clustering. Its process can be divided into two stages: (1) dividing the given set of n sample points into c initial clusters, and calculating the membership degree of each point with respect to each cluster i; (2) constantly updating the clustering centers to improve the intra-class polymerization and the interclass difference until all the clustering centers converge. FCM improves the traditional K-means in that the value u_{ik}, i.e., the membership of point x_k to the cluster centered at u_i, is a real value in the range of [0,1] instead of binary 0 or 1. The objective function is improved by using a fuzzy exponent m (large than 1 and typically set to 2) as follows:

Table 8.3 Quantitative comparison of the SOD algorithms on Group A [25]

Index	Result	Liu	Singh	BBO	RingBBO	RandBBO	EBO
Time	Average	8.110	16.809	6.944	7.035	**6.928**	6.988
	Max	8.837	46.189	7.785	8.127	7.663	7.858
	Min	7.064	8.100	5.928	5.868	5.886	5.917
	Std	0.594	13.449	0.632	0.738	0.604	0.633
Precision	Average	0.843	0.913	**0.991**	**0.991**	**0.991**	**0.991**
	Max	1.000	1.000	1.000	1.000	1.000	1.000
	Min	0.535	0.357	0.963	0.963	0.963	0.963
	Std	0.204	0.225	0.014	0.014	0.014	0.014
Recall	Average	0.896	**0.955**	0.949	0.949	0.949	0.949
	Max	0.961	0.987	1.000	1.000	1.000	1.000
	Min	0.659	0.909	0.889	0.889	0.889	0.889
	Std	0.098	0.024	0.038	0.038	0.038	0.038
F-measure	Average	0.850	0.914	**0.969**	**0.969**	**0.969**	**0.969**
	Max	0.974	0.982	0.982	0.982	0.982	0.982
	Min	0.682	0.499	0.941	0.941	0.941	0.941
	Std	0.118	0.168	0.016	0.016	0.016	0.016

Table 8.4 Quantitative comparison of the SOD algorithms on Group B [25]

Index	Result	Liu	Singh	BBO	RingBBO	RandBBO	EBO
Time	Average	8.005	18.260	6.725	6.697	**6.674**	6.687
	Max	8.383	29.197	7.452	7.360	7.184	7.117
	Min	7.782	11.760	6.310	6.192	6.260	6.339
	Std	0.271	7.081	0.431	0.425	0.345	0.294
Precision	Average	0.792	0.700	0.921	0.906	**0.930**	**0.930**
	Max	0.960	0.994	1.000	0.999	1.000	1.000
	Min	0.488	0.203	0.700	0.631	0.714	0.751
	Std	0.170	0.328	0.114	0.145	0.109	0.093
Recall	Average	0.903	**0.945**	0.924	0.919	0.911	0.918
	Max	0.980	1.000	0.989	0.991	0.990	0.992
	Min	0.784	0.765	0.800	0.781	0.764	0.779
	Std	0.077	0.090	0.067	0.078	0.087	0.079
F-measure	Average	0.837	0.750	0.916	0.897	0.914	**0.919**
	Max	0.973	0.979	0.983	0.983	0.981	0.983
	Min	0.593	0.332	0.725	0.687	0.716	0.753
	Std	0.129	0.243	0.099	0.113	0.104	0.089

Table 8.5 Quantitative comparison of the SOD algorithms on Group C [25]

Index	Result	Liu	Singh	BBO	RingBBO	RandBBO	EBO
Time	Average	8.163	15.374	6.859	**6.778**	6.787	6.786
	Max	8.434	16.366	7.182	7.180	7.167	7.127
	Min	7.901	14.607	6.365	6.420	6.403	6.450
	Std	0.277	0.737	0.351	0.328	0.329	0.304
Precision	Average	0.720	0.668	0.878	0.858	**0.879**	0.866
	Max	0.995	0.843	0.975	0.975	0.973	0.966
	Min	0.221	0.447	0.784	0.781	0.801	0.766
	Std	0.295	0.186	0.072	0.071	0.064	0.076
Recall	Average	0.904	**0.957**	0.831	0.840	0.844	0.820
	Max	0.986	1.000	0.976	0.975	0.958	0.984
	Min	0.823	0.862	0.702	0.694	0.744	0.611
	Std	0.071	0.056	0.125	0.119	0.101	0.152
F-measure	Average	0.760	0.759	0.824	0.838	**0.851**	0.831
	Max	0.900	0.896	0.913	0.896	0.898	0.904
	Min	0.359	0.582	0.749	0.731	0.776	0.675
	Std	0.228	0.130	0.071	0.074	0.059	0.096

$$J_m = \sum_{i=1}^{c} \sum_{k=1}^{n} (u_{ik})^m \, \|x_k - v_i\|^2 \tag{8.21}$$

In FCM, the cluster centers are iteratively updated as follows:

$$u_{ik} = \frac{1}{\sum_{j=1}^{c} \left(\dfrac{\|x_k - v_i\|}{\|x_k - v_j\|} \right)^{2/(m-1)}} \tag{8.22}$$

$$v_i = \frac{\sum_{k=1}^{n} x_k (u_{ik})^m}{\sum_{k=1}^{n} (u_{ik})^m} \tag{8.23}$$

FCM has been widely used because it can avoid the setting of thresholds and the intervention of the clustering by introducing the concept of membership degree of fuzzy set to image segmentation [15]. However, FCM has some shortcomings; e.g., it can be very hard to determine the number of clusters and their initial centers, and the iterative convergence process can be very time-consuming and easy to fall into local optima [8, 31]. Therefore, recently some researches combine FCM with evolutionary algorithms to simultaneously handle multiple clustering solutions to improve clustering speed and accuracy [9].

8.4.2 The Proposed Hybrid BBO-FCM Algorithm

The main disadvantage of FCM is that its performance highly depends on the quality of the initial cluster centers. In [13] Li proposes a hybrid BBO-FCM algorithm to combine the advantage of BBO with FCM for image segmentation. In the algorithm, an initial matrix of cluster centers is regarded as a solution (habitat) of BBO. The algorithm starts by initializing a population of solutions and then uses BBO to evolve the solutions to search for the optimal initial matrix, where the fitness of each solution is evaluated by employing FCM to obtain the clustering result according to the matrix. Algorithm 8.3 presents the framework of the BBO-FCM algorithm.

Algorithm 8.3: BBO-FCM algorithm

---

1 Randomly initialize a population of N solutions (initial matrices);
2 **while** *the stop criterion is not satisfied* **do**
3   **for** $i = 1$ *to* N **do**
4   Call FCM to produce the clustering result according to the ith cluster center matrix;
5   Use the reciprocal of the objective function (8.21) as the fitness of the solution;

6   Calculate the migration and mutation rates of the solutions;
7   **for** $i = 1$ *to* N **do**
8   **for** $k = 1$ *to* n **do**
9   **if** rand() $< \lambda_i$ **then**
10   Select another solution H_j with probability $\propto \mu_j$;
11   Replace the kth component (cluster center) of H_i with the corresponding one of H_j;

12   **for** $i = 1$ *to* N **do**
13   **for** $k = 1$ *to* n **do**
14   **if** rand() $< \pi_i$ **then**
15   Replace the kth component (cluster center) of H_i with a random value within the range;

16 **return** *the best solution found so far;*

In [28], we, respectively, use DE/BBO [7], localized DE/BBO (L-DE/BBO) [29], and EBO [30] to replace the original BBO in the above framework. The performance of the improved BBO-FCM algorithms is demonstrated by the experiments below.

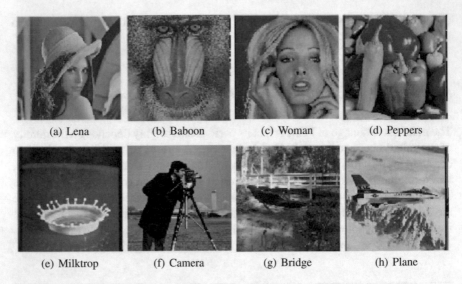

(a) Lena   (b) Baboon   (c) Woman   (d) Peppers

(e) Milktrop   (f) Camera   (g) Bridge   (h) Plane

Fig. 8.5 Eight test images for segmentation [28]

8.4.3   Computational Experiments

We compare the performance of the basic FCM, the original BBO-FCM [13], and its three improved versions [28] and three other hybrid algorithms that combine FCM with heuristics including AFSA-FCM, ABC-FCM, and PSO-FCM [9], on a set of eight images shown in Fig. 8.5. For FCM, we set the maximum number of iterations to 50, the fuzzy parameter m to 2, and the number of cluster centers to 4. The stop condition is that the difference between two adjacent solutions is less than $\zeta = 10^{-5}$ or the algorithm reaches the maximum number of iterations. The parameters of AFSA-FCM, ABC-FCM, and PSO-FCM are set as suggested in the literature. For BBO-FCM, we set the population size $N = 10$, the maximum mutation rate $\pi_{\max} = 0.005$, the number of maximum non-improving iterations $l_N = 6$ (for localized BBO), $\eta^{\max} = 0.7$, and $\eta^{\min} = 0.4$ (for EBO).

We use the following five performance metrics from [2, 4]:

- Subarea coefficient (SC), which measures the ratio of the sum of compactness and separation of the clusters:

$$SC = \sum_{i=1}^{c} \frac{\sum_{k=1}^{n}(u_{ik})^{m} \|x_k - v_i\|^2}{n_i \sum_{j=1}^{c} \|v_j - v_i\|^2} \tag{8.24}$$

where n_i is the number of samples which belongs to clustering i.

- Xie and Beni's index (XB), which measures the ratio of the total variation within clusters and the separation of clusters:

$$XB = \frac{\sum\limits_{i=1}^{c} \sum\limits_{k=1}^{n} (u_{ik})^m \|x_k - v_i\|^2}{n \min\limits_{1 \leq i \leq c, 1 \leq k \leq n} \|x_j - v_i\|^2} \tag{8.25}$$

- Classification entropy (CE), which measures the fuzziness of cluster partition:

$$CE = -\frac{1}{n} \sum\limits_{i=1}^{c} \sum\limits_{k=1}^{n} u_{ik} \log u_{ik} \tag{8.26}$$

- Separation index (S), which uses a minimum distance separation for partition validity:

$$S = \frac{\sum\limits_{i=1}^{c} \sum\limits_{k=1}^{n} (u_{ik})^2 \|x_k - v_i\|^2}{n \min\limits_{1 \leq i \leq c, 1 \leq k \leq n} \|v_j - v_i\|^2} \tag{8.27}$$

- Partition coefficient (PC), which measures the overlapping between clusters:

$$PC = \frac{1}{n} \sum\limits_{i=1}^{c} \sum\limits_{k=1}^{n} (u_{ik})^2 \tag{8.28}$$

According to their definitions, the smaller values of SC, XB, and CE or larger values of S and PC, the better the clustering performance.

Tables 8.6, 8.7, 8.8, 8.9, 8.10, 8.11, 8.12, and 8.13 present the metric values of the algorithms on the eight images, respectively. None of the algorithms can overwhelm the others in all metrics. Generally, the basic FCM has good PC values, but has bad values of the other four metric values. In particular, its XB values are very large while its S values are very small, which indicate that its partition quality is low. AFSA-FCM achieves good XB value but shows poor results in terms of the other four metrics. PSO-FCM and the original BBO-FCM achieve relatively balanced results in terms of the five metrics, and the overall performance of BBO-FCM is better than PSO-FCM. Compared with the original BBO-FCM, the three improved versions achieve better results in terms of four metrics except XB, and the improvement of EBO is the most obvious.

Figure 8.6 compares the segmented results of image Lena by FCM and the seven heuristic algorithms, from which we can see that the image obtained by EBO-FCM is better than others because its outline is more clear, especially the areas of lips and eyes. The other seven images segmented by the algorithms have similar features and are thus are omitted due to page limits (see [13, 28] for more details).

Table 8.6 Comparison of the algorithms on image Lena

Metric	FCM	AFSA-FCM	ABC-FCM	PSO-FCM	BBO FCM	DE/BBO FCM	L-DE/BBO-FCM	EBO FCM
SC	1.07	2.27	**0.57**	5.34	1.27	1.21	1.18	1.16
XB	24.04	0.35	0.90	0.39	**2.50E–01**	1.36E–07	3.60E–07	3.13E–05
CE	0.99	2.04	1.35	2.09	**8.51E–05**	5.63E–06	2.01E–06	**1.64E–06**
S	1.07E–04	0.55	0.15	**1.40**	0.28	0.46	0.61	0.61
PC	**0.53**	0.33	0.06	0.39	2.49E–11	7.29E–03	1.65E–02	0.08

Table 8.7 Comparison of the algorithms on image Baboon

Metric	FCM	AFSA-FCM	ABC-FCM	PSO-FCM	BBO FCM	DE/BBO FCM	L-DE/BBO-FCM	EBO FCM
SC	1.61	**0.96**	1.3	2.04	2.35	1.86	1.75	1.70
XB	21.46	1.74	1.74	**0.01**	0.14	0.18	0.31	0.09
CE	0.81	1.26	1.24	0.46	1.40	0.98	0.59	**0.39**
S	1.23E–04	0.30	0.39	0.61	0.67	0.75	**0.93**	0.85
PC	0.58	**1.03**	0.14	0.01	0.13	0.21	0.36	0.39

Table 8.8 Comparison of the algorithms on image Woman

Metric	FCM	AFSA-FCM	ABC-FCM	PSO-FCM	BBO FCM	DE/BBO FCM	L-DE/BBO-FCM	EBO FCM
SC	1.10	3.33	**1.03**	1.10	2.05	1.33	1.10	1.08
XB	21.36	**7.39E–09**	0.08	0.74	1.75E–04	2.69E–03	2.54E–03	0.01
CE	0.81	1.53	1.17	1.14	0.07	0.05	0.05	**0.04**
S	1.10E–04	1.26	0.25	0.27	0.64	1.01	0.98	**1.33**
PC	**0.60**	0.16	0.17	0.28	1.36E–04	0.02	0.03	0.10

Table 8.9 Comparison of the algorithms on image Peppers

Metric	FCM	AFSA-FCM	ABC-FCM	PSO-FCM	BBO FCM	DE/BBO FCM	L-DE/BBO-FCM	EBO FCM
SC	**1.10**	1.69	**2.18**	1.55	2.39	1.92	1.57	1.37
XB	40.98	0.01	**1.48E–08**	0.20	0.08	0.13	0.06	0.06
CE	0.82	0.51	**0.0018**	1.46	1.19	0.50	0.09	0.05
S	3.78E–05	0.42	0.5278	0.46	0.69	0.73	**0.88**	0.85
PC	**0.60**	0.01	8.87E–09	0.14	0.06	0.12	0.18	0.37

Table 8.10 Comparison of the algorithms on image Milktrop

Metric	FCM	AFSA-FCM	ABC-FCM	PSO-FCM	BBO FCM	DE/BBO FCM	L-DE/BBO-FCM	EBO FCM
SC	0.88	2.14	2.80	1.30	2.25	1.55	1.30	**0.73**
XB	32.08	**0.01**	0.42	0.47	0.04	0.09	0.32	0.04
CE	0.57	0.49	0.57	0.95	1.05	0.75	0.57	**0.48**
S	6.38E–05	0.74	1.08	0.29	0.72	1.03	1.02	**1.21**
PC	**0.71**	0.01	0.24	0.23	0.09	0.23	0.26	0.35

Table 8.11 Comparison of the algorithms on image Camera

Metric	FCM	AFSA-FCM	ABC-FCM	PSO-FCM	BBO FCM	DE/BBO FCM	L-DE/BBO-FCM	EBO FCM
SC	0.70	**0.27**	2.65	2.01	2.07	1.77	1.63	0.95
XB	27.64	3.96	**1.10E–03**	0.42	0.23	0.42	0.36	0.46
CE	0.44	0.13	0.55	1.46	1.35	0.55	0.22	**0.09**
S	7.08E–05	0.11	0.86	0.65	0.61	0.81	0.90	**0.93**
PC	0.78	**3.73**	1.87	0.42	0.21	0.42	0.42	0.80

Table 8.12 Comparison of the algorithms on image Bridge

Metric	FCM	AFSA-FCM	ABC-FCM	PSO-FCM	BBO FCM	DE/BBO FCM	L-DE/BBO-FCM	EBO FCM
SC	2.06	1.09	7.50	2.20	0.37	0.39	0.37	**0.35**
XB	31.13	1.63	**0.05**	0.35	4.62	4.33	8.38	8.06
CE	0.86	0.84	1.20	1.23	0.11	**0.09**	0.30	0.11
S	1.55E–04	0.35	**2.77**	0.62	0.13	0.30	0.30	0.35
PC	**0.54**	0.04	0.12	0.16	1.05E–04	0.02	0.04	0.09

Table 8.13 Comparison of the algorithms on image Plane

Metric	FCM	AFSA-FCM	ABC-FCM	PSO-FCM	BBO FCM	DE/BBO FCM	L-DE/BBO-FCM	EBO FCM
SC	1.13	0.91	1.23	1.41	1.60	1.46	1.23	**0.76**
XB	35.03	**0.27**	1.47	0.49	0.67	0.81	1.23	1.00
CE	**0.40**	0.84	0.78	1.43	1.06	0.72	0.72	0.55
S	9.84E–05	2.60E–01	0.56	0.47	0.63	0.84	0.79	**0.96**
PC	0.78	0.24	**1.40**	0.39	0.66	0.80	0.90	0.96

Generally speaking, EBO-FCM combines global migration and local migration to better balance exploration and exploitation than the other versions and thus achieves clustering results that better balance the intra-class polymerization and the interclass difference. Therefore, in most conditions we suggest using EBO-FCM except when the real-time requirement is very strict. DE/BBO-FCM is the most efficient one, because the DE operators are typically fast and the algorithm does not employ any

(a) FCM   (b) AFSA-FCM   (c) ABC-FCM   (d) PSO-FCM

(e) BBO-FCM   (f) DE/BBO-FCM   (g) L-DE/BBO-FCM   (h) EBO-FCM

Fig. 8.6 Segmented results of six algorithms on the Lena image [28]

local topology to maintain the neighborhood structure of the solutions, and thus, it is more suitable for high-performance applications.

In summary, the proposed BBO-FCM and its variants effectively segment the test images and achieve clear results, which shows the combination of BBO and FCM is beneficial to improve the accuracy and precision of image segmentation. However, a shortcoming of BBO-FCM is that its PC values are typically small, which indicates the overlapping between the resultant clusters is small. We can find that the other BBO versions cannot significantly improve the PC values, mainly due to the inherently similar migration operators they use. Thus, we do not suggest using BBO-FCM for images with many overlapping objects.

8.5 Summary

Image processing often needs to handle a huge number of pixels/regions in the image. This chapter describes the application of the BBO and its variants in image processing problems, including FIC, SOD, and image segmentation. Experimental results demonstrate the effectiveness of BBO in solving such optimization problems on a set of test images.

References

1. Alpert S, Galun M, Brandt A, Basri R (2012) Image segmentation by probabilistic bottom-up aggregation and cue integration. IEEE Trans Pattern Anal Mach Intell 34:315–26. https://doi.org/10.1109/TPAMI.2011.130
2. Balasko B, Abonyi J, Feil B (2014) Fuzzy clustering and data analysis toolbox for use with MATLAB. In: Proceedings of the IEEE fuzzy systems conference, pp 1–5
3. Batra D, Kowdle A, Parikh D, Luo J (2010) iCoseg: interactive co-segmentation with intelligent scribble guidance. In: computer vision and pattern recognition, pp 3169–3176. https://doi.org/10.1109/CVPR.2010.5540080
4. Bezdek JC (1981) Pattern recognition with fuzzy objective function algorithms. Plenum, New York
5. Borji A, Sihite DN, Itti L (2014) What/where to look next? modeling top-down visual attention in complex interactive environments. IEEE Trans Syst Man Cybern B. Cybern 44:523–538. https://doi.org/10.1109/TSMC.2013.2279715
6. Fisher Y (1995) Fractal image compression. Springer Science & Business Media, Berlin
7. Gong W, Cai Z, Ling CX (2010) DE/BBO: a hybrid differential evolution with biogeography-based optimization for global numerical optimization. Soft Comput 15:645–665. https://doi.org/10.1007/s00500-010-0591-1
8. Gong M, Liang Y, Shi J, Ma W, Ma J (2013) Fuzzy c-means clustering with local information and kernel metric for image segmentation. IEEE Trans Image Process 22:573–584. https://doi.org/10.1109/TIP.2012.2219547
9. Hruschka E, Campello R, de Castro L (2004) Evolutionary search for optimal fuzzy c-means clustering. In: Proceedings of the IEEE international conference on fuzzy systems, pp 685–690. https://doi.org/10.1109/FUZZY.2004.1375481
10. Itti L, Koch C, Niebur E (1998) A model of saliency-based visual attention for rapid scene analysis. IEEE Trans Pattern Anal Mach Intell 20:1254–1259. https://doi.org/10.1109/34.730558
11. Jacquin AE (1992) Image coding based on a fractal theory of iterated contractive image transformations. IEEE Trans Image Process 1:18–30. https://doi.org/10.1109/83.128028
12. Jing Y (2003) Research on fast algorithm for fractal image compression. Master's thesis, Southern Medical University; The First Military Medical University
13. Li Z (2013) Image segmentation technology and application based on biogeography-based optimization. Mathesis (2013)
14. Liu T, Yuan Z, Sun J, Wang J, Zheng N, Tang X, Shum HY (2011) Learning to detect a salient object. IEEE Trans Pattern Anal Mach Intell 33:353–367. https://doi.org/10.1109/TPAMI.2010.70
15. Liu L, Sun SZ, Yu H, Yue X, Zhang D (2016) A modified fuzzy c-means FCM clustering algorithm and its application on carbonate fluid identification. J Appl Geophys 129:28–35. https://doi.org/10.1016/j.jappgeo.2016.03.027
16. Ma H, Simon D (2011) Blended biogeography-based optimization for constrained optimization. Eng Appl Artif Int 24:517–525. https://doi.org/10.1016/j.engappai.2010.08.005
17. Mandelbrot BB (1977) The fractal geometry of nature. W. H. Freeman, New York
18. Navalpakkam V, Itti L (2006) An integrated model of top-down and bottom-up attention for optimizing detection speed. In: IEEE computer society conference on computer vision and pattern recognition, pp 2049–2056. https://doi.org/10.1109/CVPR.2006.54
19. Pham DL, Xu C, Prince JL (2000) Current methods in medical image segmentation. Annu Rev Biomed Eng 2:315–337. https://doi.org/10.1146/annurev.bioeng.2.1.315
20. Simon D (2008) Biogeography-based optimization. IEEE Trans Evol Comput 12:702–713. https://doi.org/10.1109/TEVC.2008.919004
21. Singh N, Arya R, Agrawal RK (2014) A novel approach to combine features for salient object detection using constrained particle swarm optimization. Pattern Recognit 47:1731–1739. https://doi.org/10.1016/j.patcog.2013.11.012
22. Talha AM, Junejo IN (2014) Dynamic scene understanding using temporal association rules. Image Vis Comput 32:1102–1116. https://doi.org/10.1016/j.imavis.2014.08.010

23. Tseng CC, Hsieh JG, Jeng JH (2008) Fractal image compression using visual-based particle swarm optimization. Image Vis Comput 26:1154–1162. https://doi.org/10.1016/j.chaos.2005.07.004
24. Wang ZC, Wu XB (2014) Hybrid biogeography-based optimization for integer programming. Sci World J 2014:9. https://doi.org/10.1155/2014/672983
25. Wang Z, Wu X (2016) Salient object detection using biogeography-based optimization to combine features. Appl Intell 45:1–17. https://doi.org/10.1007/s10489-015-0739-x
26. Wu MS, Teng WC, Jeng JH, Hsieh JG (2006) Spatial correlation genetic algorithm for fractal image compression. Chaos Solitons Fractals 28:497–510
27. Zhang W, Wu QMJ, Wang G, Yin H (2010) An adaptive computational model for salient object detection. IEEE Trans Multimed 12:300–316. https://doi.org/10.1109/TMM.2010.2047607
28. Zhang M, Jiang W, Zhou X, Xue Y, Chen S (2017) A hybrid biogeography-based optimization and fuzzy C-means algorithm for image segmentation. Soft Comput. https://doi.org/10.1007/s00500-017-2916-9
29. Zheng YJ, Ling HF, Wu XB, Xue JY (2014) Localized biogeography-based optimization. Soft Comput 18:2323–2334. https://doi.org/10.1007/s00500-013-1209-1
30. Zheng YJ, Ling HF, Xue JY (2014) Ecogeography-based optimization: enhancing biogeography-based optimization with ecogeographic barriers and differentiations. Comput Oper Res 50:115–127. https://doi.org/10.1016/j.cor.2014.04.013
31. Zhu W (2009) Image segmentation based on clustering algorithms. Ph.D. thesis, Jiangnan University

Chapter 9
Biogeography-Based Optimization in Machine Learning

Abstract Artificial neural networks (ANNs) have powerful function approximation and pattern classification capabilities, but their performance is greatly affected by structural design and parameter selection. This chapter introduces how to use BBO and its variants for optimizing structures and parameters of ANNs. The results show that BBO is a powerful method for enhancing the performance of many machine learning models.

9.1 Introduction

Since the original work of McCulloch and Pitts [12] in the 1940s, artificial neural networks (ANNs) have become an active research field in artificial intelligence and a powerful and intelligent computational tool in a variety of application areas. During the last years, deep neural networks (DNNs) have inspired a new wave of interest to and achieved great success in highly complex machine learning problems [7].

Nevertheless, the performance of ANNs is seriously affected by the selection of their structures and parameters. Traditional training methods, such as the back-propagation (BP) algorithm [5], have drawbacks including long training time, over-fitting, premature convergence. Heuristic optimization algorithms provide an effective tool for ANN design and training. This chapter introduces the application of BBO algorithms for optimizing structures and/or parameters of ANNs, including the traditional shallow forward ANNs, fuzzy neural networks, and DNNs.

9.2 BBO for ANN Parameter Optimization

9.2.1 The Problem of ANN Parameter Optimization

The most widely used ANN is the three-layer feedforward ANN, which consists of an input layer, a middle (hidden) layer, and an output layer, as shown in Fig. 9.1. Each layer contains a set of artificial neurons that can only be connected to the neurons of

© Springer Nature Singapore Pte Ltd. and Science Press, Beijing 2019   199
Y. Zheng et al., *Biogeography-Based Optimization: Algorithms and Applications*,
https://doi.org/10.1007/978-981-13-2586-1_9

Fig. 9.1 Structure of a three-layer feedforward neural network

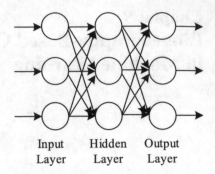

Input   Hidden   Output
Layer   Layer   Layer

the next layer, and thus the information is passed in a feedforward way. Each neuron in the input layer directly accepts an input, while each neuron i in the hidden or output layer accepts a set of inputs x_{ij} from the neurons of the previous layer and produces an output y_i as follows:

$$y_i = f \left(\sum_{j=1}^{n} w_{ij} x_{ij} - \theta_i \right) \qquad (9.1)$$

where n is the number of inputs, w_{ij} is the weight of the jth input, θ_i is the threshold of the neuron, and f is the activation function which typically uses the sigmoid function or the hyperbolic tangent function.

The numbers of neurons in the input layer and the output layer are determined by the inputs and outputs of the problem modeled by the ANN, and the number of neurons of the hidden layer is often set empirically. Once the structure of the network has been fixed, we need to tune the parameters including all connection weights w_{ij} and all neuron thresholds θ_i, such that the actual output of the ANN on the training set is as close as possible to the expected output. Suppose that the training set has N samples, the input–output pair of each sample is $(\mathbf{x}_i, \mathbf{y}_i)$, the output of the ANN for $\mathbf{x}_i$ is $\mathbf{o}_i$, the ANN training is to minimize the root mean square error (RMSE) between the actually outputs and the expected outputs:

$$\min \text{RMSE} = \sqrt{\frac{1}{N} \sum_{i=1}^{N} \|\mathbf{y}_i - \mathbf{o}_i\|^2} \qquad (9.2)$$

Thus, the ANN training problem can be regarded as a high-dimensional optimization problem. Typically, the values of weights and thresholds can be limited in the range of [0, 1] or [−1, 1]. Let n_1, n_2, and n_3 be the number of neurons in the input layer, the hidden layer, and the output layer, respectively, the problem dimension is $(n_1 n_2 + n_2 n_3 + n_1 + n_2 + n_3)$.

9.2.2 BBO Algorithms for ANN Parameter Optimization

The traditional BP algorithm [5] for ANN training is based on gradient descent. Starting from a random initial setting of network parameters, the algorithm iteratively produces an output for each training sample, calculates the error between the actual output and the expected output, and then "back-propagates" its error to adjust the network parameters such that the error is decreased along a descent direction. BP algorithm is computationally cheap, but it often suffers slow convergence rate and is easy to be trapped in local optima. In recent years, heuristic optimization algorithms such as GA and PSO have been applied to and have shown good performance in ANN training [8, 21].

Here we use BBO for the ANN parameter optimization problem, where each solution (habitat) is encoded as a n-dimensional real-value vector. For a three-layer feedforward ANN, we have

$$n = (n_1 n_2 + n_2 n_3 + n_1 + n_2 + n_3) \tag{9.3}$$

The algorithm framework is shown in Algorithm 9.1. As indicated by lines 4–6, evaluating the fitness of each solution involves constructing an ANN instance and training it through all samples, which is often computationally expensive. Thus, the population size and the maximum number of iterations of the algorithm are often limited.

Algorithm 9.1: BBO for ANN parameter optimization

1 Randomly initialize a population of solutions;
2 **while** *the stop condition is not satisfied* **do**
3   **foreach** *solution H in the population* **do**
4   Construct a network instance according to the parameters specified by H;
5   Calculate the RMSE of the network instance on the training set;
6   Evaluate the fitness of H which is reversely proportional to the RMSE;

7   Calculate the migration and mutation rates of the solutions;
8   **foreach** *solution H in the population* **do**
9   **foreach** *dimension of H* **do**
10   **if** rand() $\leq \lambda(H)$ **then**
11   Select another solution H' with a probability proportional to $\mu(H)$;
12   Perform migration from H' to H at this dimension;

13   **foreach** *solution H in the population* **do**
14   **foreach** *dimension of H* **do**
15   **if** rand() $\leq \pi(H)$ **then**
16   Perform mutation on H at this dimension;

17 **return** *the best solution found so far;*

9.2.3   Computational Experiment

We use the original BBO [16], B-BBO [11], Local-BBO using the ring topology [22], and EBO [23] to implement the above framework and use the algorithms to train an ANN for classifying three categories of wines according to 13 attributes including alcohol content, acidity, color. Accordingly, the input layer has 13 neurons, while the output layer has one neuron to output an integer value representing the type of wine. The number of neurons in the hidden layer is empirically set to 11. The ANN is trained on the WINE data set from the UCI Repository [1]. The set contains 178 samples. The experiments are divided into three groups, where the ratio of training set size to the test set size is set to 2:3, 3:2, and 4:1, respectively, in order to increase the diversity of training results. On each group, each algorithm is run for 20 times and evaluated in terms of the average classification accuracy, i.e., the percentage of the number of samples that are correctly classified. All the algorithms use the same MNFE of 1000 (i.e., the maximum number of ANN training is 1000).

The classification results of the ANN trained by the BBO algorithms are compared to those trained by GA [8] and PSO [21] on the three groups, as presented in Tables 9.1, 9.2 and 9.3, respectively.

The results show that EBO achieves the best classification results on all three groups of experiments, followed by B-BBO. On the first and second groups, EBO and B-BBO achieve the same maximum classification accuracy, but EBO has higher mean accuracy. On the second group, EBO uniquely achieves the best maximum, minimum, and mean accuracy among all the algorithms. Particularly, on the second group, the best ANN trained by EBO only misclassifies two samples among the 119 test samples; on the third group, the ANN trained by EBO only misclassifies one

Table 9.1 Classification accuracies of ANNs trained by the heuristic algorithms on the first group

Metric	GA	PSO	BBO	B-BBO	L-BBO	EBO
Max	83.18	92.52	88.79	**96.26**	90.65	**96.26**
Min	79.44	90.65	82.24	92.52	86.92	94.39
Mean	81.33	91.67	85.2	94.8	88.22	**95.13**
Std	1.88	0.9	2.12	1.79	1.35	1.02

Table 9.2 Classification accuracies of ANNs trained by the heuristic algorithms on the second group

Metric	GA	PSO	BBO	B-BBO	L-BBO	EBO
Max	84.87	93.28	89.08	96.64	91.6	**97.48**
Min	80.67	91.6	84.87	93.28	88.24	94.96
Mean	83.15	92.35	87.11	95.15	90.28	**96.29**
Std	2.06	0.89	2.33	1.63	1.72	1.15

Table 9.3 Classification accuracies of ANNs trained by the heuristic algorithms on the third group

Metric	GA	PSO	BBO	B-BBO	L-BBO	EBO
Max	83.33	94.44	88.89	**97.22**	88.89	**97.22**
Min	75	94.44	80.56	91.67	83.33	97.22
Mean	80.78	94.44	82.9	95.05	85.2	**97.22**
Std	3.2	0	3.16	2.13	1.9	0

among the 36 samples. The results show the promising ability of EBO for training the ANN for this classification problem.

The performance of GA is the worst among the six algorithms, mainly because its crossover and mutation operators often lead to premature convergence. Among the four versions of BBO algorithm, the original BBO performs the worst because its clonal-based migration greatly limits the solution diversity. In comparison, the migration operators of B-BBO and EBO are much more efficiently in exploring the solution space of the ANN parameters.

9.3 BBO for ANN Structure and Parameter Optimization

9.3.1 The Problem of ANN Structure and Parameter Optimization

The problem and algorithms studied in Sect. 9.2 assume that the ANN structure has been fixed, and thus only the connection weights and thresholds are to be optimized. However, the ANN performance also depends on its structure. For a three-layer ANN, if the number n_2 of neurons in the hidden layer is too small, the computing power of the ANN will be limited, and the training will be easily trapped in local optima; on the contrary, if n_2 is too large, the training time will be prolonged, and the training results are more likely to be over-fitting.

Therefore, it is reasonable to simultaneously optimize the structure and parameters of ANN. That is, n_2 should be an additional decision variable to be optimized. Experiences have shown that the value range of n_2 can be set as follows:

$$\log_2 n_1 \leq n_2 \leq \sqrt{n_1 + n_3} + 10 \tag{9.4}$$

where n_1 and n_3 are the number of neurons in the input layer and the output layer, respectively.

Besides the minimization of RMSE on the training set, we also expect that n_2 can be as small as possible so as to simplify the network structure. According to Eq. (9.4), the maximum number of neurons of the network is $n_1 + n_3 + \sqrt{n_1 + n_3} + 10$, and here we set the objective function as follows:

$$\min f = w \left(\frac{1}{2} \sum_{i=1}^{N} \|\mathbf{y}_i - \mathbf{o}_i\|^2 \right) + (1 - w) \frac{n_2}{n_1 + n_3 + \sqrt{n_1 + n_3} + 10} \qquad (9.5)$$

where w represents the weight of RMSE in the objective function, and its value is typically between 0.6 and 0.9.

Note that the new decision variable n_2 does not simply add a dimension to the search space of the problem. Instead, it makes the problem a variable-dimensional optimization problem. Where the value of n_2 changes, as indicated by Eq. (9.3), the number of other network parameters also changes. This needs to be treated specifically in the optimization algorithm.

9.3.2 BBO Algorithms for ANN Structure and Parameter Optimization

We use BBO to simultaneously optimize ANN structure and parameters. The main difficulty is to handle variable dimensions of different solutions in the population.

Solution encoding and initialization We use a variable-dimensional encoding for the solutions to the problem, where n_2 is the first component in the solution vector, followed by other $(n_1 n_2 + n_2 n_3 + n_1 + n_2 + n_3)$ parameters. When initializing a population, instead of randomly setting a value of n_2 for each solution, we assign all possible values of n_2 evenly to all solutions in the population. For example, for the classification problem of the WINE data set, according to Eq. (9.4) n_2 is in the range of [3,14] and thus has 12 possible values. Suppose the population size is 50, then each value is assigned to at least four solutions, respectively (the remaining two solutions will be assigned with two random values of n_2). Let K be the number of possible values of n_2, we set the population size N as

$$N = \max(50, 4K) \qquad (9.6)$$

Consequently, in the population each solution has at least three other solutions that have the same dimension. However, when the data set has too many features, K will be very large, and thus the above initialization strategy can be inapplicable. Thus for very high-dimensional data sets, we normally limit that $K \leq 125$ and hence $N \leq 500$.

Migration For the variable-dimensional problem, the migration from an emigrating solution H' to an immigrating solution H can be divided into three cases.

Case 1 H' and H have the same dimension, for which the BBO migration operator can be directly applied.

Case 2   The dimension of H' is larger than that of H. Let $n_2 = H(1)$ be the number of hidden neurons in the ANN represented by H and $n'_2 = H'(1)$, we first make $H(1)$ immigrate from $H'(1)$ to determine the length of the H after migration, and then make each remaining component of H immigrate from the corresponding component of H'.

Case 3   The dimension of H' is smaller than that of H. We first migrate $H'(1)$ to $H(1)$ to determine the length of the H after migration and then migrate the remaining components of H' to H, and finally randomly set the rest undetermined components of H.

Mutation The mutation operation is also first applied to $n_2 = H(1)$ to determine the length of the H after migration. Here we employ a Gaussian mutation as:

$$n_2 = \text{norm}(n_2, K/4) \tag{9.7}$$

The result n_2 will be rounded to the nearest integer and, if it exceeds the range specified by Eq. (9.4), will be set to the nearest boundary. Afterward, each remaining component is mutated. If there are undetermined components, they will be randomly set within the predefined range.

Algorithm Framework and Implementation The algorithm framework for ANN structure and parameter optimization based on B-BBO is shown in Algorithm 9.2. We respectively use the original BBO [16], B-BBO [11], and EBO [23] to implement the framework. For BBO, an immigrating solution H is always set as the same length as the emigrating solution H'. For B-BBO and the local migration of EBO, the length of the solution after migration is between its original length and the length of the emigrating solution. For the global migration of EBO, let H' and H'' be the neighboring and non-neighboring emigrating solutions of H, the length of H after migration is between $[\min(|H'|, |H''|), \max(|H'|, |H''|)]$.

9.3.3 Computational Experiment

Here, we also compare the proposed BBO algorithms with GA [8] and PSO [21] for training ANN on the WINE data set from the UCI Repository [1]. Except the population size, the parameters of algorithms are set the same as that in Sect. 9.2. As the problem of structure and parameter optimization is more complex, the MNFE is set to 2500.

The experiments are also divided into three groups as described in Sect. 9.2.3. Tables 9.4, 9.5, and 9.6 present the maximum, minimum, average, and standard

---

Algorithm 9.2: BBO for ANN structure and parameter optimization

---

1 Randomly initialize a population of N solutions, where N is set based on Eq. (9.6);
2 **while** *the stop condition is not satisfied* **do**
3   **foreach** *solution H in the population* **do**
4   Construct a network instance according to the parameters specified by H;
5   Calculate the RMSE of the network instance on the training set;
6   Evaluate the fitness of H based on Eq. (9.5);

7   Calculate the migration and mutation rates of the solutions;
8   **foreach** *solution H in the population* **do**
9   **if** rand() $\leq \lambda(H)$ **then**
10   Select another solution H' with a probability proportional to $\mu(H)$;
11   Migrate $H'(1)$ to $H(1)$;
12   Let $\widehat{k} = \min(H(1), |H'|)$;
13   **for** $k = 2$ *to* $\widehat{k}$ **do**
14   Migrate $H'(k)$ to $H(k)$;
15   **if** $\widehat{k} < H(1)$ **then**
16   **for** $k = \widehat{k} + 1$ *to* $H(1)$ **do**
17   Randomly set $H(k)$ in the range as Eq. (9.4);

18   **foreach** *solution H in the population* **do**
19   **if** rand() $\leq \pi(H)$ **then**
20   Set $H(1)$ based on Eq. (9.7);
21   **for** $k = 2$ *to* $H(1)$ **do**
22   Randomly mutate $H(k)$ in the range as Eq. (9.4);

23 **return** *the best solution found so far;*

Table 9.4 Classification accuracies of ANNs trained by the algorithms on the first group

Metric	GA	PSO	BBO	B-BBO	EBO
Max	84.11	94.39	86.92	**96.26**	**96.26**
Min	79.44	93.46	83.18	95.33	96.26
Mean	81.70	93.69	84.89	95.55	**96.26**
Std	2.01	0.71	1.61	0.95	0

deviation of the classification accuracy of the ANNs trained by the algorithms on the three groups, respectively.

The results show that B-BBO and EBO achieve much better classification accuracies than the other three algorithms. The performance of GA is the worst because its crossover and mutation operators often lead to premature convergence. The original BBO performs better than GA but worse than PSO, mainly because its clonal-based migration operator limits the solution diversity to a great extent. But by using the blended migration and the combination of local and global migration, B-BBO and EBO show great performance improvement.

Table 9.5 Classification accuracies of ANNs trained by the algorithms on the second group

Metric	GA	PSO	BBO	B-BBO	EBO
Max	84.87	94.12	88.24	**97.48**	**97.48**
Min	82.35	90.76	85.71	95.80	97.48
Mean	83.60	92.23	86.86	96.61	**97.48**
Std	1.12	1.55	1.22	0.91	0

Table 9.6 Classification accuracies of ANNs trained by the algorithms on the third group

Metric	GA	PSO	BBO	B-BBO	EBO
Max	80.56	97.22	88.89	**100**	**100**
Min	77.78	97.22	77.78	100	100
Mean	79.32	97.22	81.38	100	100
Std	2.05	0	5.16	0	0

B-BBO and EBO obtain the exact same results on the third group, i.e., they both obtain an ANN that can make correct classification on all samples of the data set. On the first and second groups, B-BBO and EBO obtain the same maximum classification accuracy, but EBO achieves a better mean classification accuracy than B-BBO, which demonstrates the performance improvement brought by neighborhood structure and the combination of two migration operators of EBO.

Comparing the experimental results in this section and in the previous section, we can see that for GA and BBO, the classification accuracies obtained by optimizing both network structure and parameters are even worse than those are obtained by only optimizing parameters. This is because the inclusion of the structure parameter greatly enlarges the solution space (14 times), and GA and BBO cannot efficiently explore such a large solution space. The other three algorithms show much better performance as they can find better network structures (if exist) than the structure set based on experience.

9.4   BBO for Fuzzy Neural Network Training

9.4.1   The Problem of FNN Training

Fuzzy logic and ANN are two widely used tools in intelligent computing [13]. ANN is suitable for extracting complex functions from large data sets, but is not good at expressing domain knowledge such as rules. On the contrary, fuzzy logic is good at expressing and reasoning knowledge, but its learning ability is limited. Fuzzy neural network (FNN) is a computational model that combines fuzzy logic and neural network for tacking with machine learning problems with uncertain and intuitive domain knowledge.

Figure 9.2a presents a typical structure of FNN [10] that consists of the following five layers from the bottom to the top (where use $u_i^{(l)}$ denote the ith input to the lth layer, and $a_i^{(l)}$ to denote the corresponding output).

Layer 1   The input layer, where the nodes directly transmit inputs to the next layer:

$$a_i^{(1)} = x_i \tag{9.8}$$

Layer 2   The fuzzification layer, where each node represents a fuzzy set defined by a Gaussian membership function and calculates the output as

$$a_{ir}^{(2)} = \exp\left(-\frac{(u_i^{(2)} - \mu_{ir})^2}{2\sigma_{ir}^2}\right) \tag{9.9}$$

where μ_{ir} and σ_{ir} are respectively the center and the width of the Gaussian function of the rth term of input $u_i^{(2)}$.

Layer 3   The fuzzy rule layer, where each node uses a prod function to match the antecedents of fuzzy rules:

$$a_r^{(3)} = \prod_{i=1}^{n} u_i^{(3)} \tag{9.10}$$

Layer 4   The normalization layer, which each node forms a linear combination of the firing strengths from Layer 3:

$$a_r^{(4)} = u_r^{(4)} \left(k_{0r} + \sum_{i=1}^{n} k_{ir} x_i\right) \tag{9.11}$$

Layer 5   The defuzzification layer, which produces the output by integrating all the actions from Layer 3 and Layer 4:

$$a^{(5)} = \frac{\sum_r u_r^{(5)}}{\sum_r u_r^{(4)}} = \frac{\sum_r \left(u_r^{(4)} \left(k_{0r} + \sum_{i=1}^{n} k_{ir} x_i\right)\right)}{\sum_r u_r^{(4)}} \tag{9.12}$$

Consequently, the FNN can represent fuzzy rules with the following form:

$$\text{Rule } r : \text{IF } (x_1 \text{ is } A_{1r}) \wedge \cdots \wedge (x_n \text{ is } A_{nr})$$
$$\text{THEN } f_r = k_{0r} + k_{1r} x_1 + \cdots + k_{nr} x_n \tag{9.13}$$

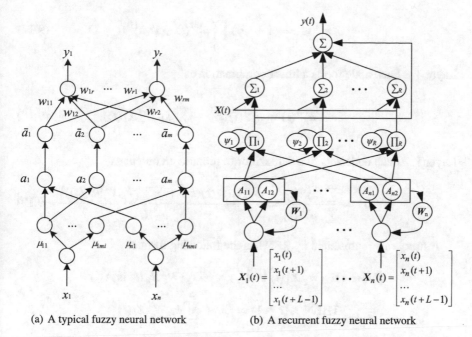

(a) A typical fuzzy neural network   (b) A recurrent fuzzy neural network

Fig. 9.2 Illustration of fuzzy neural network structures

The basic FNN can only accept static inputs. To deal with temporal inputs, we can use recurrent fuzzy network (RFN) which can store a series of historical inputs [20]. Figure 9.2b presents a RFN structure that consists of the following five layers.

Layer 1   Each node transforms a sequence of L input variables into a vector:

$$X_i(t) = [x_i(t), x_i(t+1), \ldots, x_i(t+L-1)]^T \qquad (9.14)$$

and then applies a weight vector $W_i = [w_{i1}, w_{i2}, \ldots, w_{iL}]^T$ to generate the output as

$$a_i^{(1)}(t) = W_i^T X_i(t) \qquad (9.15)$$

Layer 2   Each node represents a fuzzy set and calculates the output as

$$a_{ij}^{(2)}(t) = \exp\left(- \frac{\left(u_i^{(2)}(t) - \mu_{ij}\right)^2}{2\sigma_{ij}^2} \right) \qquad (9.16)$$

Layer 3   Each node uses fuzzy AND operator to perform antecedent matching of the fuzzy rules and uses a feedback node ψ_r to generate the output as

$$a_r^{(3)}(t) = (1 - \psi_r) \prod_{i=1}^{n} u_i^{(3)}(t) + \psi_r a_r^{(3)}(t - 1) \qquad (9.17)$$

Layer 4   Each node forms a linear combination as

$$a_r^{(4)}(t) = u_r^{(4)}(t)(k_{0r} + \sum_{i=1}^{n} k_{ir} W_i^{\mathrm{T}} X_i(t)) \qquad (9.18)$$

Layer 5   Each node acts as a defuzzifier to produce an output as

$$a^{(5)}(t) = \frac{\sum_r u_r^{(5)}(t)}{\sum_r u_r^{(4)}(t)} = \frac{\sum_r \left(u_r^{(4)}(t)(k_{0r} + \sum_{i=1}^{n} k_{ir} W_i^{\mathrm{T}} X_i(t)) \right)}{\sum_r u_r^{(4)}(t)} \qquad (9.19)$$

A fuzzy rule represented by RFN has the following format:

$$\text{Rule } r : \text{IF } (W_1^{\mathrm{T}} X_1(t) \text{ is } A_{1r}) \wedge \cdots \wedge (W_n^{\mathrm{T}} X_n(t) \text{ is } A_{nr})$$

$$\text{THEN } f_r(t + 1) = k_{0r} + \sum_{i=1}^{n} k_{ir} W_i^{\mathrm{T}} X_i(t) \qquad (9.20)$$

Here we consider a problem of optimizing RFN parameters, which include

- The Gaussian parameters $\{\mu_{11}, \sigma_{11}, \ldots, \mu_{ir}, \sigma_{ir}, \ldots, \mu_{nR}, \sigma_{nR}\}$ in Layer 2.
- The feedback weights $\{w_{11}, \ldots, w_{il}, \ldots w_{nL}\}$ used in Layer 2.
- The feedback $\{\psi_1, \ldots, \psi_r \ldots, \psi_R\}$ weights used in Layer 3.
- The connection weights $\{k_{01}, \ldots, k_{ir}, \ldots, k_{nR}\}$ used in Layer 4.

Thus a solution to the problem is a $(3nR + nL + 2R)$-dimensional vector.

9.4.2   An EBO Algorithm for FNN Parameter Optimization

After testing the original BBO [16] and its several variants, we propose a new EBO algorithm [24] shown in Algorithm 9.3 for training the RFN, which differs from the original EBO [23] in that it uses the following single form for global migration:

$$H(d) = H_{\text{far}}(d) + \alpha(H_{\text{nb}}(d) - H(d)) \qquad (9.21)$$

where $\alpha = \text{norm}(0.5, 0.2)$ is a Gaussian random number.

Algorithm 9.3: The EBO algorithm for RFN parameter optimization

1 Randomly initialize a population of solutions, each being a $(3nR + nL + 2R)$-dimensional real-value vector;
2 **while** *the stop condition is not satisfied* **do**
3   **foreach** *solution H in the population* **do**
4   Construct a RFN instance according to the parameters specified by H;
5   Calculate the RMSE of the RFN instance on the training set;
6   Evaluate the fitness of H which is reversely proportional to the RMSE;
7   Calculate the migration rates of the solutions, and update η based on Eq. (4.3);
8   **foreach** *solution H in the population* **do**
9   **foreach** *dimension d in H* **do**
10   **if** rand() $< \lambda(H)$ **then**
11   Select a neighboring solution H_{nb} with probability $\propto \mu$;
12   **if** rand() $< \eta$ **then**
13   Select a neighboring solution H_{far} with probability $\propto \mu$;
14   Perform a global migration according to Eq. (9.21);
15   **else**
16   Perform a local migration according to Eq. (4.1);
17   **if** *H is better than the original solution* **then**
18   Replace the original solution with H;

19 **return** *the best solution found so far*;

Table 9.7 Summary of the six groups of the data tuples for population classification

Group	Tuples	Area (km^2)	Crowd density (1/km^2)
#1	68	300 ∼ 500	≥1000
#2	960	250 ∼ 500	300 ∼ 1000
#3	716	200 ∼ 300	300 ∼ 1000
#4	358	100 ∼ 200	500 ∼ 1000
#5	312	100 ∼ 200	200 ∼ 500
#6	816	≤100	500 ∼ 1000

9.4.3 Computational Experiment

In [24], we use a RFN for classifying a crowd of people based on their temporary location data. The data set contains 3230 data tuples, which can be divided into six groups according to their features summarized in Table 9.7.

We compare the EBO algorithm with three other algorithms including the comprehensive learning PSO (CLPSO) [9], the multispecies PSO (MSPSO) [6], and the self-adaptive DE (SaDE) [15]. All the algorithms use the same population size of 50 and MNFE of 150000.

Table 9.8 Classification accuracies of RFNs trained by the algorithms. ©[2014] IEEE. Reprinted, with permission, from Ref. [24]

Group	CLPSO	MSPSO	SaDE	EBO
#1	76.47%+	76.47%+	77.94%+	**80.88%**
#2	68.75%+	68.33%+	70.00%+	**71.46%**
#3	74.72%+	72.35%+	75.28%	**75.56%**
#4	79.05%	77.65%+	78.89%	**81.01%**
#5	76.28%+	75.64%+	76.60%+	**80.45%**
#6	59.07%+	56.37%+	63.24%+	**68.75%**
Total	69.66%+	68.11%+	71.36%+	**73.81%**

The test uses a fivefold cross-validation, and each algorithm is run for 30 times in each cross-validation. Table 9.8 presents the average classification accuracy achieved by the algorithms over the 150 runs. In columns 2–4, a superscript $^+$ indicates that the result of EBO is significantly better than the corresponding algorithm (at 95% confidence level) according to the paired t test.

The results show that EBO exhibits the best performance among the four algorithms on each group of the training set. In terms of classification accuracy, the result of EBO is approximately 6% higher than CLPSO, 8.36% than MSPSO, and 3.43% than SaDE. Particularly, the accuracy achieved by EBO is larger than 70% on Group #1–#5 and is below 70% only on the most complex Group #6. According to the statistical test, the overall accuracy of EBO is significantly better than all the other algorithms.

Figure 9.3 presents the convergence curves of the four algorithms for RFN training. As we can see, all the methods have rapid convergence speeds over the first 15000~20000 NFEs. However, CLPSO and MSPSO get trapped in local optima at RMSE around 0.039 and 0.055, respectively. SaDE converges slightly faster than EBO in early stages, but its speed becomes slower in later stages. In comparison, EBO is capable of jumping out the local optima and continuing to explore the search space until it reaches training RMSE $= 0.027$ after about 35000 NFEs. This demonstrates that EBO has a good balance between exploration and exploitation and can effectively avoid premature convergence.

9.5   BBO for Deep Neural Network Optimization

9.5.1   The Problem of DNN Training

Theoretically, a three-layer feedforward ANN can approximate any nonlinear continuous function to an arbitrary accuracy [8]. Nevertheless, when the problem dimension becomes very large, the performance of traditional ANNs decreases dramatically.

Fig. 9.3 Convergence curves of the four algorithms for RFN training. ©[2014] IEEE. Reprinted, with permission, from Ref. [24]

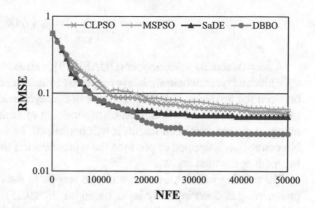

Recent development of DNNs, i.e., ANNs using multiple hidden layers, provides a powerful tool for modeling high-dimensional and complex systems. As a DNN can have a huge number of weights and thresholds to be optimized, a common way is to reduce the dimensionality using layer-by-layer pre-training [4], such that only a limited number of parameters are optimized at a time.

Autoencoder is one of the most common building blocks of DNN which attempts to reconstruct its input $\mathbf{x} \in [0, 1]^n$ at its output [3]. It first transforms (encods) its input to a hidden representation $\mathbf{y} \in [0, 1]^n$ through an affine mapping:

$$f_\theta(\mathbf{x}) = s(\mathbf{W}\mathbf{x} + \mathbf{b}) \tag{9.22}$$

where $\theta = [\mathbf{W}, \mathbf{b}]$, $\mathbf{W}$ is a $n' \times n$ weight matrix and $\mathbf{b}$ is a n'-dimensional bias vector. And $\mathbf{y}$ is then mapped back (decoded) to a reconstructed vector $\mathbf{z} \in [0, 1]^n$ in input space with appropriately sized parameters $\theta' = [\mathbf{W}', \mathbf{b}']$:

$$g_{\theta'}(\mathbf{y}) = s(\mathbf{W}'\mathbf{y} + \mathbf{b}') \tag{9.23}$$

The training of autoencoder is to minimize the average reconstruction error over the training set D:

$$\arg \min_{\theta, \theta'} = \frac{1}{|D|} \sum_{\mathbf{x} \in D} \|\mathbf{x}, g_{\theta'}(f_\theta(\mathbf{x}))\|^2 \tag{9.24}$$

A denoising autoencoder (DAE) is a variant of the basic autoencoder to reconstruct a clean "repaired" input from a corrupted one [18]. That is, a DAE first corrupts an initial input $\mathbf{x}$ into $\widetilde{\mathbf{x}}$ with stochastic noise, and then maps the corrupted input $\widetilde{\mathbf{x}}$, as with the basic autoencoder, to a hidden representation $\mathbf{y} = f_\theta(\widetilde{\mathbf{x}}) = s(\mathbf{W}\widetilde{\mathbf{x}} + \mathbf{b})$ from which we reconstruct a $\mathbf{z} = g_{\theta'}(\mathbf{y}) = s(\mathbf{W}'\mathbf{y} + \mathbf{b}')$.

The training of DAE is still to minimize the average reconstruction error, but the key difference is that $\mathbf{z}$ is now a deterministic function of $\widetilde{\mathbf{x}}$ rather than $\mathbf{x}$ and thus the result of a stochastic mapping of $\mathbf{x}$:

$$\arg\min_{\theta,\theta'} = \frac{1}{|D|} \sum_{\mathbf{x} \in D} \|\mathbf{x}, g_{\theta'}(f_\theta(\widetilde{\mathbf{x}}))\|^2 \qquad (9.25)$$

A deep denoising autoencoder (DDAE) [19] is an extension of DAE that has multiple hidden layers, where each layer captures complicated, higher-order correlations between the activities of hidden features in the layer below. Note that input corruption is only used for the initial denoising-training of each individual layer; once the mapping f_θ has thus been learnt, it will henceforth be used on uncorrupted inputs; No corruption is applied to produce the representation that will serve as clean input for training the next layer.

The training of DDAE consists of two stages. In the first stage, an unsupervised pre-training is done layer by layer based on Eq. (9.25). In the second stage, a supervised training is done on the whole network to minimize the RMSE between the actual outputs and expected outputs.

9.5.2 An EBO Algorithm for DNN Structure and Parameter Optimization

Classical gradient-based methods for DNN training still have drawbacks such as premature convergence and difficulty in determining the number of hidden neurons. We propose an EBO algorithm for both unsupervised pre-training and supervised training for DDAE [17].

For unsupervised pre-training, the algorithm is applied to optimize the DDAE layer by layer. When optimizing the lth hidden layer, a solution is encoded as a variable-size vector $W = \{\mathbf{w}_1, \ldots, \mathbf{w}_i, \ldots, \mathbf{w}_n\}$, where n is the number of neurons in the current layer. The local and global migration operators defined in Sect. 4.3 are employed in the algorithm here, while the following strategies are used for dealing with variable dimensions (where η is the immaturity index):

- For global migration, if the dimension n' of one emigrating solution is larger than the dimension n'' of another emigrating solution, then after the n'' dimension the global migration is degraded to local migration between W and the first emigrating solution.
- If the dimension n of W is smaller than the dimension n' of the emigrating solution W', then each sub-vector $\mathbf{w}'_i$ in W' has a probability of η of being cloned to W ($n < i \le n'$).
- If the dimension n is larger than n', then each sub-vector $\mathbf{w}_i$ in W has a probability of $(1 - \eta)$ of being removed ($n' < i \le n$).

For supervised learning, since the structure of the DDAE is fixed, the algorithm is directly applied to optimize the network parameters as a whole.

Table 9.9 Average prediction percentage errors of the DDAEs trained by the algorithms. Reprinted from Ref. [17], © 2016, with permission from Elsevier

Metric	Greedy	BBO	GA	HS	EBO
Max	218.56%	230.77%	192.08%	166.28%	116.84%
Min	1.55%	0.59%	1.96%	1.75%	1.02%
Mean	39.30%	29.26%	30.59%	21.47%	21.16%
Std	0.333	0.294	0.293	0.228	0.177

9.5.3 Computational Experiment

We use the proposed EBO algorithm to train a DDAE for predicting morbidity of gastrointestinal infections by food contamination [17]. 227 types of contaminants and 119 types of food are selected, and the input dimension is 8426 (many input variables can be missing or noisy). The data set comes from four counties in central China from August 2015 to July 2016 (50 weeks), and thus the number of data tuples is 200. We compare EBO with other four algorithms, including the gradient-based algorithm, the original BBO [16], an improved GA [2], and an improved HS algorithm [14], for DDAE training.

Table 9.9 presents average prediction percentage errors of the DDAEs trained by the algorithms. Among the five algorithms, the average error rate achieved by the greedy algorithm is much higher than that of the other heuristic optimization algorithms, which indicates that the gradient-based algorithm is easy to be trapped in local optima, while the heuristic algorithms evolve populations of solutions to simultaneously explore multiple regions in the solution space and thus are much more capable of jumping out of local optima.

Among the four heuristic algorithms, EBO achieves the lowest average error. BBO and GA exhibit similar performance that is much worse than EBO, which shows that the combination of local and global migration in EBO is much more effective than the original clonal-based migration. HS also exhibits a good ability in exploring the solution space for DDAE structure and parameter optimization, and it achieves an average error close to DDAE-EBO. However, the number of cases whose error rates are less than a small threshold (here set to 15%) achieved by HS is much lower than that of EBO, which indicates that EBO performs better in fine-tuning the DDAE and thus reaches high prediction accuracies ($\geq 85\%$) on more tuples than HS.

9.6 Summary

This chapter introduces the application of BBO and its improved version for neural network training which involves essentially high-dimensional optimization problems. A distinct feature of the problem is that the fitness evaluation is often very expensive, as it needs to construct a network instance based on the parameters represented by the solution and then test the network instance on the whole training set. Thus, the efficiency of the algorithm is of critical importance. The experimental results demonstrate the efficacy of BBO-based heuristic algorithms in machine learning.

References

1. Blake CL, Merz CJ (1998) UCI repository of machine learning databases. http://www.ics.uci.edu/~mlearn/MLRepository.html
2. David OE, Greental I (2014) Genetic algorithms for evolving deep neural networks. In: Proceedings of the GECCO, pp 1451–1452. https://doi.org/10.1145/2598394.2602287
3. Hinton GE (1989) Connectionist learning procedures. Artif Intell 40(1):185–234. https://doi.org/10.1016/0004-3702(89)90049-0
4. Hinton GE, Salakhutdinov RR (2006) Reducing the dimensionality of data with neural networks. Science 313(5786):504–507
5. Hopfield JJ (1982) Neural networks and physical systems with emergent collective computational abilities. Proc Nat Acad Sci 79:2554–2558. https://doi.org/10.1073/pnas.79.8.2554
6. Juang CF, Hsiao CM, Hsu CH (2010) Hierarchical cluster-based multispecies particle-swarm optimization for fuzzy-system optimization. IEEE Trans Fuzzy Syst 18:14–26. https://doi.org/10.1109/TFUZZ.2009.2034529
7. Lee H, Grosse R, Ranganath R, Ng AY (2009) Convolutional deep belief networks for scalable unsupervised learning of hierarchical representations. In: Proceedings of the 26th annual international conference on machine learning, pp 609–616. https://doi.org/10.1145/1553374.1553453
8. Leung FHF, Lam HK, Ling SH, Tam PKS (2003) Tuning of the structure and parameters of a neural network using an improved genetic algorithm. IEEE Trans Neural Netw 14:79–88. https://doi.org/10.1109/TNN.2002.804317
9. Liang JJ, Qin AK, Suganthan P, Baskar S (2006) Comprehensive learning particle swarm optimizer for global optimization of multimodal functions. IEEE Trans Evol Comput 10:281–295. https://doi.org/10.1109/TEVC.2005.857610
10. Lin CT (1996) Neural fuzzy systems: a neuro-fuzzy synergism to intelligent systems. Prentice Hall PTR, Upper Saddle River
11. Ma H, Simon D (2011) Blended biogeography-based optimization for constrained optimization. Eng Appl Artif Int 24:517–525. https://doi.org/10.1016/j.engappai.2010.08.005
12. McCulloch WS, Pitts WH (1943) A logical calculus for the ideas immanent in nervous activity. Bull Math Biophys 5:115–133. https://doi.org/10.1007/BF02478825
13. Negnevitsky M (2005) Artificial intelligence: a guide to intelligent systems. Pearson Education, Essex (2005)
14. ao Paulo Papa J, Scheirer W, Cox DD (2016) Fine-tuning deep belief networks using harmony search. Appl Soft Comput 46:875–885. https://doi.org/10.1016/j.asoc.2015.08.043
15. Qin AK, Huang VL, Suganthan P (2009) Differential evolution algorithm with strategy adaptation for global numerical optimization. IEEE Trans Evol Comput 13:398–417. https://doi.org/10.1109/TEVC.2008.927706
16. Simon D (2008) Biogeography-based optimization. IEEE Trans Evol Comput 12:702–713. https://doi.org/10.1109/TEVC.2008.919004
17. Song Q, Zheng YJ, Xue Y, Sheng WG, Zhao MR (2017) An evolutionary deep neural network for predicting morbidity of gastrointestinal infections by food contamination. Neurocomputing 226:16–22. https://doi.org/10.1016/j.neucom.2016.11.018

18. Vincent P, Larochelle H, Bengio Y, Manzagol PA (2008) Extracting and composing robust features with denoising autoencoders. In: Proceedings of the 25th international conference on machine learning, pp 1096–1103. ACM, New York, NY, USA. https://doi.org/10.1145/1390156.1390294

19. Vincent P, Larochelle H, Lajoie I, Bengio Y, Manzagol PA (2010) Stacked denoising autoencoders: learning useful representations in a deep network with a local denoising criterion. J Mach Learn Res 11:3371–3408

20. Wu GD, Zhu ZW, Huang PH (2011) A TS-type maximizing-discriminability-based recurrent fuzzy network for classification problems. IEEE Trans Fuzzy Syst 19:339–352. https://doi.org/10.1109/TFUZZ.2010.2098879

21. Zhang JR, Zhang J, Lok TM, Lyu MR (2007) A hybrid particle swarm optimization-back-propagation algorithm for feedforward neural network training. Appl Math Comput 185:1026–1037. https://doi.org/10.1016/j.amc.2006.07.025

22. Zheng YJ, Ling HF, Wu XB, Xue JY (2014) Localized biogeography-based optimization. Soft Comput 18:2323–2334. https://doi.org/10.1007/s00500-013-1209-1

23. Zheng YJ, Ling HF, Xue JY (2014) Ecogeography-based optimization: enhancing biogeography-based optimization with ecogeographic barriers and differentiations. Comput Oper Res 50:115–127. https://doi.org/10.1016/j.cor.2014.04.013

24. Zheng YJ, Ling HF, Chen SY, Xue JY (2015) A hybrid neuro-fuzzy network based on differential biogeography-based optimization for online population classification in earthquakes. IEEE Trans Fuzzy Syst 23:1070–1083. https://doi.org/10.1109/TFUZZ.2014.2337938

Index

© Springer Nature Singapore Pte Ltd. and Science Press, Beijing 2019
Y. Zheng et al., *Biogeography-Based Optimization: Algorithms and Applications*,
https://doi.org/10.1007/978-981-13-2586-1

Printed in the United States
By Bookmasters